DATELINE FORT BOWIE

Charles F. Lummis.

DATELINE FORT BOWIE

Charles Fletcher Lummis Reports on an Apache War

Edited, Annotated, and with an Introduction by
Dan L. Thrapp

UNIVERSITY OF OKLAHOMA PRESS: NORMAN

By Dan L. Thrapp

Al Sieber, Chief of Scouts (Norman, 1964)
The Conquest of Apacheria (Norman, 1967)
General Crook and The Sierra Madre Adventure (Norman, 1972)
Juh: An Incredible Indian (El Paso, 1973)
Victorio and the Mimbres Apaches (Norman, 1974)
(editor) *A Cavalryman in Indian Country* (Ashland, Oregon, 1974)
Dateline Fort Bowie: Charles Fletcher Lummis Reports on an Apache War (Norman, 1979)

Library of Congress Cataloging in Publication Data

Lummis, Charles Fletcher, 1859–1928.
　　Dateline Fort Bowie.

　　Contains the entire collection of dispatches
filed by the author from the Arizona front and
published by the Los Angeles Times in 1886.
　　Includes bibliographical references and index.
　　1.　Apache Indians—Wars, 1883–1886.
2.　Fort Bowie, Ariz.　I.　Thrapp, Dan L.
II.　Title.
E83.88.L85　1979　　　973.8'4　　　78–58091
ISBN 0–8061–1494–0

For JAN—
and she knows why!

CONTENTS

LIST OF ILLUSTRATIONS

MAPS

DATELINE FORT BOWIE

INTRODUCTION

BY EARLY SPRING of 1886 it commenced to appear that the interminable Apache troubles of the Southwest were about over, in contrast to the gloomy opening of the year. Then Geronimo had depredated still. The dread war leader of the hostile Chiricahuas had gone out for yet another time back in May of 1885, and seemed on January 1 as active as ever. Bands of his fighters ravaged Sonora and either had raided into the United States or, it was feared, would cross the border at any moment. As Nana, Chatto and others had clearly demonstrated, they posed no idle threat; they could wreak havoc far exceeding what would be thought possible for numbers so few and organization so haphazard, and everyone in the Border Country feared their return. The exposed—and wise—settlers knew it might prove a catastrophe. The press was full of reminders such as this one, from a source listed only as "Exchange":

> The Apache can conceal his swart body amid the green grass, behind brown shrubs or gray rocks, with so much address and judgment that any but the experienced would pass him by without detection at the distance of three or four yards. Sometimes they will so resemble a granite boulder as to be passed within near range without suspicion. They will plant themselves among the yuccas, and so closely imitate the appearance of that tree as to pass for one of its species.

Southwesterners by and large were exasperated by the protracted military campaigns which seemed so indecisive, and their mood often surfaced as resentment against the hard-working soldiers who were primarily charged with defeating and bringing in the hostiles. Yet not all the voices were of condemnation. Some editors with military experience or understanding were more tolerant of the tediousness of the operations and the failure of tough Army men to bring a solution more promptly. For example, on January 13, 1886, the *Los Angeles Times* commented editorially, its voice no doubt that of brevet Lieutenant Colonel Harrison Gray Otis, editor, part owner, and proud veteran of the Civil War:

3

It has become rather popular on this Coast to leap with agility and violence upon the soldierly frame of General George Crook [in command of the military Department of Arizona], the best Indian fighter in the United States Army, as we have always contended, because he has not conducted his campaigns in a manner to please the able editorial Jominis,[1] the profound Territorial statesmen, and the patriotic contractors of Arizona. . . .

Several wise editors in their sanctums, and one or two Congressmen between drinks, have made the announcement that "General Crook is a great failure," and they point to "a score of settlers who have been killed by the murderous Apaches" to prove their assertions. Such people are poor readers of history who imagine that these warlike savages, the craftiest and bravest of their kind, are going to sit down in camp and let white men come up and lead them away captive. A campaign against the savage Apaches over the cactus fields and almost impassable mountains of Arizona is attended with difficulties which few people understand who have never been over the ground. . . .

And the war went on. So far as most southwesterners could see on New Year's Day, it might continue forever.

On the eleventh of that month (although word of the tragedy did not reach the U.S. press until January 28) Captain Emmet Crawford, a capable and universally esteemed officer of great Indian-fighting experience, had been shot and mortally wounded by Mexican irregulars in the Sierra Madre at the very point of opening negotiations with the hostiles whom he had tracked relentless to this most secluded hiding place. When word of the sad event became known, a solution to the stubborn "Apache problem" must have seemed even more remote than before. Yet Crawford's prodigious work among the little-known mountains of Mexico was not entirely in vain. His subordinate, First Lieutenant Marion Maus, extricating the command from this ticklish situation, finally reached the border with old Nana himself, one other warrior, and four women. Furthermore, he brought the pledge of "the surrender of the remainder of the hostiles at the full of the next moon." The moon turned full on February 19 in 1886; the Apaches did not arrive by that date for their decisive talks with Crook, but they did not miss the following full moon (March 19) by very much. Of course, having no written calendar, they timed their activities by lunar phases or other natural phenomena.

Rumors spread, "infallibly" reporting the movements of the elusive hostiles during the months of waiting. Flack increased over the Crawford shooting, which threatened to become an international incident of some moment.

---

[1]Antoine Henry Jomini (1779–1869) was a Swiss general who served with Napoleon until after a clash with a fellow member of Bonaparte's staff, when he went over to the Russians and became a celebrated authority on strategy and war theorist.

Seeking to avert this possibility, Mexican officials opened serious investigations into the tragedy, while Mexican publications sought to justify the slaying by tales of depredations by "lawless" and uncontrollable Apache scouts under purely nominal American supervision. There was a report from Chihuahua that:

> The Mexicans . . . are much excited. . . . Statements are circulated to the effect that the United States will demand redress and invade Chihuahua to avenge the killing. . . . [Chihuahuans] all declared that the killing of Crawford was accidental and was unfortunate, and believe that the United States is simply anxious for a pretext to overrun their country. . . .

U.S. Army men, hopeful that Maus's report of the imminent surrender of the hostiles would be borne out, prepared to implement any Apache decision to come in, ignoring so far as possible such gratuitous advice as this from the *New York Times* of February 13:

> It is said that [Crook] will adopt more stringent measures than ever to keep [the Indians] on the reservation. . . . Can it be that these butchers, who have since May last slain from 150 to 200 settlers in cold-blood, are to be placed once more on the reservation and allowed to roam about on it as they roamed before? We hope not. They should be punished by imprisonment at least. It never should be possible for them to repeat their bloody work.

A program aiming at imprisonment or other harsh punishment, as Crook and other veterans knew, would have made surrender of the hostiles unlikely in the extreme, besides being perhaps unjust.

Crook passed through Tucson from his headquarters at Prescott's Fort Whipple on February 23, enroute for Fort Bowie to await arrival of the hostiles at some point where a talk might be had. He did not leave Bowie until a month later, on March 23, for the San Bernardino ranch in the southeast corner of Arizona. From there he would drop south of the Line for the critical meeting and council. But before he left Bowie, he took the occasion of a communication from Tucson to give the lie to persistent news reports of war developments.

Such dispatches often were based upon imagination, rather than events. They had alleged, for example, a Crook-Geronimo talk as early as February 25, or thereabouts; again they had revealed a "hot skirmish" between Apaches and Mexicans on March 19, reporting that the hostiles had fled north and surrendered unconditionally to someone. There was little novelty about such fictitious revelations, which were very numerous.

The southwestern press, and newspapers in the remainder of the country, for that matter, were filled with accounts of "Apache war" developments, but unless they came from official or otherwise reliable sources, they were hardly trustworthy. No newspaper except for local sheets of scant circulation, and no wire service had a staff man anywhere near the military center of operations, Fort Bowie, or with any of the several commands working the mountains of Arizona, New Mexico, Sonora or Chihuahua. Telegraphers at the various way stations picked up what rumors were circulating and sometimes filed them to news outlets for payment at so much a line. It might reasonably be supposed that on dull days, with little stirring, a few individuals in need of pocket change dreamed up notions of what might be happening somewhere, and sent it to El Paso or Tucson as the "latest from the front." James J. Chatham, a newsman who had worked at Tucson and Tombstone (and who would publish papers in Nogales and Cananea), spoke with some authority on these matters. He informed the *Citizen:* "To one who knows something of the situation of Indian affairs, the newspaper articles which have been published by the Tombstone, Tucson, and other papers assume more the dignity of mirth than truth. In Tombstone you can hear any kind of Indian story you want. It is furnished you with your coffee and eggs, and interest is added by the great complicity of age between the news and the eggs. I am very sorry to say, however, that in every case the eggs rank."

The Army sometimes took cognizance of the more outrageous yarns published, and occasionally even issued a denial or clarification. Just before he left Bowie, therefore, Crook took time out to wire a Tucson photographer, Willis P. Haynes, who had sought accurate information on the status of hostilities, perhaps with an eye to profiting with his camera as Tombstone's Camillus Fly would actually do at the forthcoming council. Crook's message said:

> There isn't a word of truth in the reports in the Associated Press with reference to my [already] having had an interview with the hostiles, or that Geronimo has surrendered, or that the hostiles were in Lieut. Maus's camp when it was attacked by the Mexicans . . . , or in the many similar reports which have been so persistently circulated.
>
> Lieut. Maus, with his battalion of scouts, is in camp south of San Bernardino ranch, and is in communication with the hostiles, who are also in camp in his vicinity, waiting for a conference with me. I leave Fort Bowie to-morrow morning for this purpose, and until I have a talk with the hostiles it is mere silliness to predict the result.

The lack of on-the-spot news coverage was not the fault of the Army nor in all probability of the General.

Crook might have recalled a mutually profitable relationship with John F. Finerty of the Chicago *Times* and other newsmen in 1876 during his northern Plains operations against the Sioux. In 1883 he had taken A. Frank Randall, better remembered as a photographer but a sometimes correspondent too, on his great Sierra Madre campaign, although the results of that experiment with a reporter are not clear. If Randall filed dispatches anywhere upon conclusion of the adventure, the fact is not incontrovertibly known. Yet Crook had found no reason to fear or be wary of competent reporters. Despite his often repeated disclaimer of interest in derogatory articles about him and his work, he was as solicitous of his image as anyone else. He might even have wished at times for factual lay descriptions of his work to be spread for public view, but there had been no one to do this—until Charles Fletcher Lummis thought of it and easily sold the idea to his boss, Colonel Otis of the *Los Angeles Times,* who was said to have served for a time with Crook during the Civil War. Martial-minded Otis, later to become a brevet Major General during the unpleasantness with Spain, would have listened sympathetically, beyond a doubt.

Lummis, with his boundless energy, already was whacking out a unique niche in southwestern life and culture. His life and accomplishments have been fully recorded by Dudley Gordon[2] and others, and need be only summarized here.

Born at Lynn, Massachusetts, March 1, 1859, Charles Fletcher Lummis was educated at Harvard, and his first real literary effort, *Birch Bark Poems,* was written while he was a student and printed on bark Lummis may have collected himself. In 1884 he hiked from Ohio to Los Angeles, sending weekly dispatches describing what he saw, learned, experienced, and thought about, to the Los Angeles newspaper which carried them for a delighted readership. The stories so intrigued Otis that he personally rode a buggy from Los Angeles out to nearby San Gabriel to meet the walker on his arrival, relating the incident to his paper's readers in the issue of February 3, 1885. Lummis had covered 3,507 miles in 143 days and had reported things no one else ever had or would report, since Lummis saw life differently from anyone else. He was a character, but one gifted with a brilliant, ever-curious, ever-inquiring intellect, a man with no prejudices that could not be easily lifted, and a freethinker in a Victorian period when freethinking and freeacting were rarities. In appearance, with his rumpled clothing and wildly-floating aura of stubborn hair, Lummis might have seemed a precursor to the so-called "youth culture" of our own day, but there was more depth and more warmth to him. He was something of a scientist in his clear and meticulous thought, and his views were original and generally sound. He would later become a city editor of *The Times,* remaining with the paper for two or three years in all.[3]

Afterward his career progressed at a varied though rapid pace and in several

---

[2]Dudley Gordon, *Charles F. Lummis: Crusader in Corduroy* (Los Angeles, Cultural Assets Press, 1972).

[3]Lummis joined *The Times* following his arrival at Los Angeles on February 2, 1885, becoming its entire local news staff. After nineteen months he bought into the paper and on September 10, 1886, was named city editor. He left the newspaper because of ill health late in 1887.—Gordon, *Charles F. Lummis,* 113–17.

directions. He lived for five years in the New Mexico Indian pueblo of Isleta, traveled widely through the Southwest, and visited Mexico and Peru. He founded *Out West* magazine, editing it for some years, was Los Angeles city librarian for five years, founded the Landmarks Club to preserve historic sites within California, did much to ameliorate the lot of the state's Indians, founded the widely-respected Southwest Museum in 1907, was a member and innovative force in many archaeological, literary, and historical organizations, and published a number of books, one of them, *A Tramp Across the Continent* (1902) relating his experiences walking across America, and another, *A Bronco Pegasus* (1928), including a ballad of Geronimo.[4] Lummis died November 25, 1928, full of honors and accomplishments. Most of his deeds were highly worthy, although, since he was a well-rounded man, others were a bit unorthodox for his time.

Adventure called strongly to him in this year of 1886, with Apaches on the loose and nothing very accurate being written about the campaign against them for the public prints. He had been for more than a year tied, if loosely, to an editorial desk, and he needed a change, or thought he did. Also, he figured this wild Indian business could use a dash of honest reporting; he talked about it with Otis—many times, perhaps. With the publisher's gruff blessing (an editorial paragraph March 27 read: "There is another report out that the Apache hostiles have surrendered again. The thing is growing monotonous, and we propose to find out the facts about it."), he gathered a satchel full of necessaries, including his notebooks and pencils. On March 30, *The Times* informed its morning readers:

> Chas. F. Lummis of the TIMES staff, left yesterday for Geronimo's stamping ground in Sonora, Mexico, twenty-five miles southeast of San Bernardino Springs. Sunday's telegraphic advices were that the Apache chief was holding alternately a pow-wow with his braves and a conference with General Crook, with the alternative of conditional surrender or a fight to the death. "Lum" will write directly from General Crooks's camp, and will furnish the TIMES with a series of letters and special dispatches which will throw a good deal of light on the much-vexed Indian question, and will keep our readers well informed as to events. Should the campaign continue he will probably remain in the field with General Crook's forces as long as his legs hold out and his scalp sticks fast.

But it was not quite all that cut and dried. A late bulletin on another page of the same issue of the paper reported (ominously, from Lummis's viewpoint, no doubt):

[4]Charles F. Lummis, *A Bronco Pegasus* (New York, Houghton Mifflin Company, 1928). The Geronimo poem, "Man-Who-Yawns," is on pp. 33–43, followed by a prose exposition, pp. 43–50.

FORT BOWIE, A.T., March 29.—News is received today that the four Apache chiefs. Geronimo, Chihuahua, Nana and Natchez with twenty-nine bucks and forty-eight squaws unconditionally surrendered to General Crook . . . on Saturday last. The captives were placed in charge of Lieut. Maus, who is now conducting them to this point.

The reporter had gone too far with his preparations to back out now, however. After a no doubt frenzied appeal to Otis, Lummis scored his point and *The Times* added to its earlier editorial note:

LATER—Since the above was written, telegraphic information has been received of the surrender of Geronimo. Mr. Lummis will, however, proceed to the scene of the late difficulties, and will detail the many points of interest which a man of his industry and love of adventure can gather.

The eager writer sped to the train before the boss could change his mind, as newspaper executives have been known to do. The Great Adventure was on.

It will be observed that his dispatches are of two principal species: long, complete articles that Lummis filed by mail for use at the paper's pleasure, and brief news-oriented items doubtless sent by wire, in the manner of any reporter on the spot who had the resourcefulness—and the resources—to hire someone to take them horseback to the telegrapher's office, many miles removed from Fort Bowie and Apache Pass, in the tiny community of Bowie on the route through Railroad Pass. In addition, while the first stories bear datelines and no doubt were sent from Arizona, the latter articles are not so identified, and must have been written from his notes after his return to Los Angeles, which was well within a month of his setting out. The volume of material he collected on the brief expedition was impressive; truly he was a hard-working reporter. But then Lummis ever was a man of diligence and limitless energy.

Lummis reveals sound judgment in his reporting, maturing within a very short time into a remarkable understanding of the basic facts of the military situation as they are accepted today by virtually all historical writers who have assessed them fairly. But today's writers have the advantage of a century of perspective. Lummis did not have that. This fact makes the soundness of his judgments all the more creditable, and speaks for his perception and objectivity.

His work is perhaps most valuable today for the fresh insights it gives of the personalities involved and for on-the-spot scenes, such as the unforgettable picture he draws of Crook's conference with Chihuahua and other Apaches in the General's office. It is a delineation that could never have been recovered otherwise, not even from Bourke, for that intelligent and literate officer had too much intimate experience with the figures and routine of frontier life to see with quite the same freshness what appeared novel to Lummis. Lummis wrote what he saw, and we are in his debt for faithfully and fully recording it.

He also gives word pictures of characters who are all but overlooked by the historical record: of Frank Bennett, for instance, and Santiago McKinn, who aside from Lummis and a passing mention by Bourke would have been but a name in the dusty archives. Thanks to this reporter he becomes a living, breathing, bawling witness to Apache treatment of their numerous child-captives.

While the reporter obtained much of his material from formal reports that still are available to us, he was a good enough newsman to flesh them out with what he could discover firsthand, assisted as he was by a quick curiosity, balanced judgment and his endless industry.

Among the most engaging of his dispatches are those which take us through the thickets of life on the perimeter of the grim business of Indian war. Such are his revealing articles on the bombastic cowboys imbued with a "let me at 'em!" spirit toward Apaches which rarely survived emergence from the saloon. Of the would-be scouts such as Buckskin Frank Leslie, Lummis concedes he may have been a fine fellow, but one with whom truth has room to roam and "as much of it as you could pick up on the point of a pin would last him for a year." Too, there is Lummis's touching tour of the Fort Bowie graveyard, with its lonely wooden markers bearing dour legends and revealing tales no other reporter has collected to give us, summarizing, as they do, the grimmest aspect of the business of conquering the frontier.

If there remains much that Lummis omitted, there is still a great deal that he *did* write that makes us duly appreciative. His exuberant, free-wheeling style gives an impression of flamboyancy and inaccuracy that close examination of the dispatches generally dispels. There is basic accuracy here, and one may with some confidence depend upon the truthfulness of his reports, as regards the nouns if not quite all of the adjectives. Lummis was always Lummis, and when he began a sentence no one could predict, not even Lummis himself, quite where it was going to come out. Yet Lummis was ever the quality observer and writer—and he was always honest. It should be pointed out, and here is as good a place as any to do so, that there is a lack of consistency in the spelling of some proper names, particularly of Indians, and of such nouns as bronco, which occasionally appears as broncho. These lapses may be the fault not of Lummis but of some copy reader or telegrapher and should cause little confusion.

There has been some rearrangement of the order in which his dispatches appeared in the newspaper. This was done to make the book more cohesive and orderly, and to tie related items together. Since his wired dispatches often interrupted the mailed items, sometimes at awkward points, these have been shifted occasionally in a minor way for purposes of forming a logical sequence. But they are all here, exactly as written. Every word identifiable as penned by

Lummis for *The Times* from the Arizona front is here included, with the dates of publication and, when datelined, the times filed. It makes a unique record. Other concurrent dispatches not by Lummis are sometimes inserted, either when Lummis later rebuts them or for purposes of clarification of the record, for the benefit of the reader.

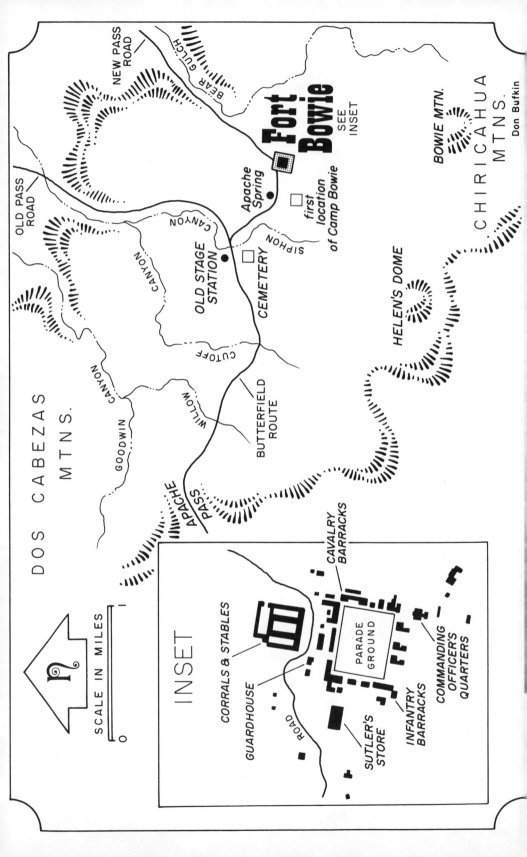

DOS CABEZAS MTNS.

NEW PASS ROAD

BEAR GULCH

OLD PASS ROAD

Fort Bowie

SEE INSET

Apache Spring

first location of Camp Bowie

CANYON

SIPHON CANYON

BOWIE MTN.

Don Bufkin

CHIRICAHUA MTNS.

OLD STAGE STATION

CEMETERY

CANYON

GOODWIN CANYON

WILLOW CANYON

CUTOFF

BUTTERFIELD ROUTE

HELEN'S DOME

APACHE PASS

N

SCALE IN MILES

0

INSET

CORRALS & STABLES

CAVALRY BARRACKS

GUARDHOUSE

ROAD

PARADE GROUND

COMMANDING OFFICER'S QUARTERS

SUTLER'S STORE

INFANTRY BARRACKS

CHAPTER I

Los Angeles Times, April 3, 1886:

THE VISITING M. C.'S

And What They Said on Their Journey Eastward.

AT THE SCENE OF INDIAN TROUBLES.

*The Hostiles and Their Surrender—Gov. Zulick and
Ex.-Gov. McCook—Indian Campaigning in a Pullman Car.*

[STAFF CORRESPONDENCE OF THE TIMES.]

BOWIE, A.T., March 30.—Congressmen are first rate in their way—and I won't be mean enough to say that their way should be a long way off. They are always ornamental, and occasionally useful. The ornamental part you have already sampled without my telling, and as for the usefulness, I don't think Los Angeles will be left behind any procession in which the members of the present committee have any grand-marshal finger. Of course it was expected that our guests should praise us as long as they were within gunshot—Congressmen are not necessarily devoid of horse sense. When Mrs. Jones comes to see us she falls on the neck of the mud-frescoed children, and would even feel obliged to declare that Bose's fleas were the most graceful, polished and lovable fleas she ever saw. But wait till she gets

OUT OF THE HOUSE

and meets Mrs. Miggs on the corner—then don't we just catch it? I always bank on what they say after their escape. Well, by that standard, Los Angeles didn't suffer. The ears of our Gracious Lady may well have burned, yesterday and to-day, at the countless clever things that have been said of her charms by the critical gentlemen upon whom she shook down such a rain of blossoms.

When Monday noon's express pulled out of Los Angeles it carried Senator John P. Jones,[1] an Angeleno favorite. He had just come down from the north, and stopped only long enough to get away with a good dinner. If it is not already a "chestnut" to you, I may relate that you

OWE THE COMMITTEE'S VISIT[2]

to Senator Jones. He wanted them to see a city in whose praise he is always apt;

---

[1] John Percival Jones, born in 1829 in England but raised in Ohio, became a U.S. Senator from Nevada in 1873 and served there continuously until 1903. He had been a sheriff of Trinity County, California, and had mining interests in California and Nevada. His affection for California, as described by Lummis, was genuine: after retirement he settled at Santa Monica and died at Los Angeles in 1912.—*Biographical Dictionary of the American Congress 1774–1971* (Washington, Government Printing Office, 1971), hereafter: *BDAC*.

[2] The legislators with whom Lummis journeyed from Los Angeles to Bowie, Arizona, were of

but they thought they couldn't stop. In fact, not knowing what they would miss, they decided not to stop. But Bre'r Jones outgeneraled them—and now they are glad of it. He found, at the last minute, that he would have to stay in 'Frisco a day longer, and they must wait for him in Los Angeles until he could catch up. And even so it came to pass.

WHAT JONES DID NOT KNOW.

En route to San Gabriel I had a little chat with the Senator. He is looking well, and says he feels so-so. To leading questions about his supposed railroad interests, he answered that he had just come across from the East with the Congressional Committee, and during his stay on the coast had been too busy with private business to look or even think of railroad matters. He knew nothing new of the talked-about road up the coast; and even expressed surprise when told of the new Santa Monica and Los Angeles railroad now being surveyed.

When the Los Angeles delegation loomed up on the San Gabriel horizon, loaded with Congressmen and contentment, our peaceful train experienced a revolution. After the boxes of oranges, cases of wine and cases of statesmanship had been shoved aboard; after Washingtonians had given three rousing cheers for Southern California and Southern California ditto for the Washingtonians, there was no more peace for the wicked. The able representatives piled into their special Pullman car, "Buckingham," happy as a big sunflower. They were full of

SCENERY, CLIMATE AND THINGS.

[James A.] Louttit, of Stockton,[3] was the sole dissenter. He was very jovial, but when they all tooted loudly for Los Angeles, the sap of the Northern Citrus Belt thawed out. He maintained up and down that his district is the best in the State, and the only one fit to tie to. Down in Los Angeles the natives raise

---

a special committee named March 12, 1886, at Washington, D.C., to accompany to California the remains of the late Senator John Franklin Miller, who had died at the Capitol on March 8 after a legislative and military career. Born November 21, 1831, Miller practiced law in his native Indiana, migrated to California, and returned to Indiana to hold political office and be commissioned an officer of volunteers during the Civil War, from which he emerged a brevet Major General. He returned to California, where he was elected to various political offices, culminating with that of U.S. Senator, March 1, 1881. He was buried at San Francisco (reinterred May 5, 1913, in Arlington National Cemetery), and the delegation from Congress which had attended the funeral was now en route back to Washington.—*BDAC; Congressional Record,* March 12, 1886: 2,297, 2,351.

[3]Born at New Orleans in 1848 but raised in California, Louttit served a single term, 1885–87, in the House of Representatives. He died at Pacific Grove, California, in 1906.— *BDAC.*

oranges to sell, but up in the N.C.B.—one of whom Stockton is which—they just raise 'em for shade trees. Wait till they get their trees up there grafted, Los Angeles will want to crawl into her burrow and retract the burrow! "Oh, if I could have got you fellows out to *my* place, I'd have shown you a welcome that you haven't seen in the State. We'd make them all ashamed of themselves!" Such was the tenor of his songs. I labored with Louttit, humanely trying to show him that while Stockton may raise very ripe insane asylums, she shouldn't stray outside her natural products. But Louttit is a shoulder-striker, with an arm of generous girth and firmness; and, as he keeps himself in trim by steady boxing in Washington, it didn't seem considerate to push even the truth after he got excited about it.

[Joseph] McKenna, of Suisun,[4] is the exact reverse of his California comrade, being quiet, sandy, of

PRESBYTERIAN BARBERING,

and very virtuous aside from a propensity to promulgate fatigued puns. He couldn't say anything too good of Los Angeles and its suburbs, and wearied Louttit.

[Polk] Laffoon,[5] of Kentucky, was the blue-grass gentleman all through—not given to bragging, but sure to get away with his companions in any of their bouts. Of course he is a horse-lover, and of course he was charmed with California's equine capabilities. After every rhapsody over Los Angeles, he would wind up with "Oh, if we only had that climate in Kentucky, what horses we could raise!" He cherished in his purse a

JETTY LOCK OF HAIR—

not the culling of some soft-eyed senorita's tresses, but some wisps which he had plucked with his own hand from the tails of the Arab and St. Julien.[6] He is going to have it made into a watchchain. Among his many friendly remarks

---

[4]McKenna, born in Philadelphia in 1843, reached Benicia, California, in 1855 and was elected to the House in 1885, serving until 1892. He became U.S. Attorney General under McKinley from 1897 until 1898, then associate justice of the U.S. Supreme Court until his resignation in 1925. He died in Washington, D.C., the next year.—*BDAC*.

[5]Laffon (or Laffoon) was born in 1844 near Madisonville, Kentucky, and served the Confederacy during the Civil War. He was captured at Donelson in 1862, exchanged at Vicksburg a few months later, served with John Hunt Morgan during that cavalry commander's 1863 raid north of the Ohio River, and was captured again at Cheshire, Ohio. He was held for the remainder of the war. He served in the House from 1885–89 and died at Madisonville in 1906.—*BDAC*.

[6]Arab and St. Julien were noted trotters of the day, although Arab, "the wonder California trotter," later became better known as a road horse in the famous stable of John Shepard of Boston. Arab was described as "off to a brush [a brief race at top speed] like a rocket, steady,

were: "I wouldn't have missed this trip to Los Angeles for anything. It was a perfect revelation to me. It is a wonderful country, and a wonderful climate. Anything will grow, anywhere, if you only give it water. By the way, as one of the Committee on Public Lands, I have reported favorably on a bill to turn water from the Colorado river into part of the Colorado desert."[7]

[John T.] Spriggs[8] was as far from a sprig as possible, and his rotund figure and monkish face were always to be found in the thickest of the fray, a personalization of earth-shaking jollity.

[Seth L.] Milliken, of Maine,[9] and [William P.] Hepburn, of Iowa,[10] were the quietest of the outfit, and they didn't let anything get away from them, either.

ONE OF WALKER'S PIRATES.

Louttit had a sociable friend along, a Mr. Adams, of Sacramento, who was one of [William] Walker's Nicaragua State-stealers,[11] and related some interesting anecdotes of the famous expedition. When we got out to a little station west of Colton he jumped off and got a lot of the white flower-stalks of the mescal, which he kept on the car all night. He said it reminded him of the time, after the gang had captured Mazatlan, taken the Governor and other big-bugs prisoners and sailed with them to Enseñada, the prisoners took advantage of a time while the freebooters were on shore. They bribed the mate of the brig, with promises of gold galore, to let them loose. He did so, and they sailed back

---

game and true," with few equals. St. Julien, a great Orange County, California, horse, in 1879 set a record for a mile of two minutes, 12 3/4 seconds.—Hall of Fame of the Trotter, Goshen, New York; Roger Longrigg, *The History of Horse Racing* (New York, Stein & Day, 1972), 246.

[7]Irrigation of parts of the Imperial Valley of California from the Colorado River waters had been urged as early as the 1850's by Oliver Wozencraft, a one-time Indian agent and California state legislator. Sentiment gradually had grown in the state and in Congress for funds to develop the idea, although it did not flower until after the turn of the century, receiving its greatest thrust when the river burst its banks in 1904–1905, causing floods that created today's Salton Sea.—Erwin Cooper, *Aqueduct Empire: A Guide to Water in California . . .* (Glendale, Calif., Arthur H. Clark Co., 1968).

[8]Born in England in 1825, Spriggs was raised in New York State and was a onetime mayor of Utica. He served in the House of Representatives from 1883–87 and died at Utica in 1888.—*BDAC*.

[9]Seth Milliken, born in Maine in 1831, served in the House from 1883 until his death at Washington, D.C., in 1897.—*BDAC*.

[10]Hepburn was born in Ohio in 1833, but moved to Iowa with his parents eight years later and was an officer in the Civil War with the 2nd Iowa Cavalry. He served in the House from 1881–87 and from 1893–1909. He died at Clarinda, Iowa, in 1916.—*BDAC*.

[11]Adams is not further identified. William Walker, one of America's noted filibusterers, "occupied" Nicaragua in 1855–56, and later declared himself president of that country. He was executed at Trujillo, Honduras, in 1860. His book on the Nicaraguan war does not mention Adams, but many hundreds of Americans, mostly unidentifiable today, took part in that adventure.

to Mazatlan in triumph, while the bold buccaneers, thus reft of arms, provender and plunder, danced wildly about on the shore. The ex-prisoners handsomely rewarded the treacherous mate by sending him in irons to the City of Mexico, where he shortly executed a fandango with space for a floor and a rope for a partner. The deserted buccaneers had to forage on the country for awhile, and lived on a sober diet of roasted mescal root and a pint of corn a day.

SOME MORE.

The life of the party was Col. [Charles N.] Johnson, Chief Clerk of the Senate,[12] who lent his gravity to chaperone the giddy young things from the House. As a chaperone he loomed up as an unmitigated success, and none of the party could get along without his aid.

F. B. Loomis, a correspondent of the *Philadelphia Press,* who accompanies the party to write their wrongs, is a sociable and pleasant young journalist, and made himself a convenient fixture.

Oh, yes, and there was Lieutenant Clover the gancé of the dead Senator's daughter.[13] He had least to say of any of them, and properly kept a little aloof from the madding throng.

We rolled along into the San Gorgonio Pass with gathering night, the Congressional legs getting stretched at every step. If I had a phonograph along, you should hear the able official and semi-official remarks that were passed on each place and the country at large. It is probably fortunate, however, that their hopes of re-election are free from that blight. The simple statement that we all finally got to bed is enough.

Morning found us grunting up steep grades amid the curious southern columns and candelabras of the giant cactus. Then down again into barer deserts, relieved only by their growth of tough mesquite. This has not been quite so lively a day as yesterday. There was considerable matutinal trouble with the respective hats, which had shrunken in the penetrating atmosphere of the desert. There were several incidents, but none worthy of note until we reached somnolent, diffuse and antediluvian Tucson. There we found the confirmation of good news hinted at in a private telegram received by Senator Jones at Yuma at midnight. Geronimo, the gory Chiricahua

HAD SURRENDERED

with all his murderous band, including his lieutenants, Natchez, Nana and Chihuahua—105 in all, counting bucks, squaws and papooses. The surrender took place in Sonora, about 25 miles south of the Arizona line, on Saturday;

---

[12]Johnson was chief clerk of the U.S. Senate from Dec. 18, 1883 until August 14, 1893.—Information from Francis R. Valeo, Secretary of the Senate, April 9, 1976.

[13]This reference to the friend of Senator Miller's daughter is unclear. "Lieutenant Clover" is not identified.

but the first news reached Tucson late Monday night. We found Gov. [Conrad M.] Zulick at the depot ready to start to Fort Bowie to lend [Brigadier] Gen. [George] Crook a little of his valuable advice. Zulick has been so Apacheously scalped by the Territorial press for his action in calling for United States troops for the alleged protection of the San Carlos reservation from howling citizens, that he

JUMPS AT THE CHANCE

to wipe out the score. The feeling in Tucson, so far as I could gauge it in thirty minutes, is in favor of having Geronimo and the other ringleaders turned over to the civil authorities; and it is the belief that Zulick has come up to make a formal demand for them from Gen. Crook. I tried to pump the Governor as to his intentions, but he wouldn't give anything away to-night. He had been committeed for a couple of hours then, too, and if ever his jaw runs loose, it should have been then. The 'key' was there. Besides Zulick, Judge [William H.] Barnes, of the First Judicial District of Arizona; United States Marshal [W.K.] Meade; Judge Alex. Campbell and his wife, Mrs. Barnes, and ex-Gov. [Edward M.] McCook, of Colorado, boarded our train at Tucson. Before leaving that ungalvanized municipal corpse, also, I met Ed. Stowell, late of *Pomona Progress,* who has been out here a fortnight, lung-hunting. He goes on to El Paso to-morrow. Sheriff [Robert M.] Paul, about whom such a contest is going on in the Superior Court of Pima county, and J.A. Muir, the popular Superintendent of that division of the S[outhern] P[acific] R.R., also joined the gang.

The inside of the Congressional car

GOT TOO HIGH

in temperature after that time, and I evacuated. The piles of cases of wine, brandy and other tokens of Sunny Slope, which had almost completely blocked the front and back platforms, and the passage-way became quite transcendable. McCook had fought out this and all subsequent Indian campaigns, and had obliterated the troublesome race from the countenance of the globe, long before this. He had a remarkably effective Pullman policy.

I had expected to get down to the Mexican line in time for the surrender, or at least to intercept Gen. Crook on his way back to Fort Bowie. The developments at Tucson, however, turned my nose to Bowie, the station for Fort Bowie, and I left my compañero S.W. Strong, of Los Angeles, at Benson, where he started down the Guaymas road to the big ranch of the Erie Cattle Company, in which he is interested.

CAMPED AT BOWIE.

We reached Bowie several hours late, and the Congressmen and Tucsonians embraced each other in affectionate farewell. The stage had gone to the fort, and the only remaining scheme was an ambulance, which Governor Zulick's party crammed to the point of spilling. I got in with the officers in command here—there are 300 troops encamped beside the track—and found it unnecessary to break for the fort (16 miles) in the dense darkness. Gen. Crook, by traveling with relays, got to the fort last night. [First] Lieut. [Marion P.] Maus, [14] with the troops and the prisoners, will not arrive before to-morrow. The unconditional surrender of the hostiles is a grand feather in Gen. Crook's chapeau, and a sad snub to the paper warriors who have been trying to "teach their grandmother."

I am quartered for the night in the tent of [First] Lieut. [David N.] McDonald, [15] in command of the cavalry and Indian scouts here. In the morning, Uncle Sam will horse me, and I shall get to the fort early. Meanwhile the slow-stepped sentry calls sleepily before the tent, "Twelve o'clock 'n' all's well."

CHARLES F. LUMMIS.

Los Angeles Times, March 31, 1886:

ESCAPED

Geronimo Gives His Captors the Slip.

SPECIAL DISPATCH TO THE TIMES.

BENSON, Ariz., March 30.—The Congressional Committee was joined at Tucson this afternoon by Gov. Zulick, Gen. McCook of Colorado, Judge Barnes of the First Judicial District, U.S. Marshal Meade and Judge Alex Campbell. The Tucson delegation is bound for Fort Bowie, where Gen. Crook is expected to arrive to-morrow with Geronimo, Natchez, Chihuahua, Nana and their entire following, 105 in all, who surrendered Saturday in Sonora.

---

[14]Born in Maryland, Maus (1850–1930) was a West Point graduate who had fought against the Sioux and Nez Percé before being posted to the Apache theatre. He was to receive a Medal of Honor for extricating the Sierra Madre command after the slaying of Captain Emmet Crawford in January of 1886. Later Maus took part in the Wounded Knee action of 1890 and went on to become a Brigadier General, retiring in 1909. He died at New Windsor, Maryland.—Dan L. Thrapp, *Dictionary of Frontier Characters,* manuscript in preparation.

[15]McDonald was a West Pointer from Tennessee who had had much southwestern border experience with the 4th Cavalry. In January, 1882, commanding scouts, he had crossed the Border into Mexico and was arrested there, his action becoming a minor incident in U.S.—Mexican relations. In April of that year he had narrowly survived a Juh-Loco ambush in the Stein's Peak range between Arizona and New Mexico. He resigned in 1888 and returned to Tennessee.—Dan L. Thrapp, *The Conquest of Apacheria* (Norman, University of Oklahoma Press, 1967); *General Crook and the Sierra Madre Adventure* (Norman, University of Oklahoma Press, 1972).

Gov. Zulick will hold a serious consultation with Gen. Crook, and it is believed that he will make a requisition for Geronimo and other ring-leaders to have them turned over to the civil authorities for trial. Gov. Zulick declines to outline his programme at present. The Tucsonites have given the Congressional Committee a severe earache on the Indian question. The general belief is that the ringleaders will be executed. This morning's Tucson Star says that Gen. Crook and his policy are fully vindicated.

LUM.

LATER

Geronimo and Twenty Bucks Make Their Escape

ASSOCIATED PRESS DISPATCHES TO THE TIMES.

WILLCOX, A.T., March 30.—General Crook arrived at Fort Bowie last night, leaving Lieutenant Maus in charge of Geronimo and the other surrendered Apaches. News, however, has just been received that Geronimo, with twenty Indians and some squaws, escaped during the night. Lieutenant Maus, with all the men he can spare, has started in pursuit. It is believed that Geronimo has gone to join Chief Mangus.

Los Angeles Times, April 1, 1886:

GERONIMO

Much Fire Water the Cause of His Flight.

SPECIAL DISPATCH TO THE TIMES.

FORT BOWIE, A.T., March 31.—The ill news reached here late last night, by courier from Lieut. Maus, that on the night of the 29th, Geronimo, Natchez, and nineteen other bucks left Lieut. Maus's camp during a heavy rainstorm, with their weapons but no horses. They took thirteen squaws along, and are probably still in the mountains. [Second] Lieut. [Samuel L.] Faison[16] is on his way here with the rest of the captives, and Maus is in pursuit of the fugitives.

The whole trouble was caused by an American named Tribolett,[17] who has a store 400 yards south of the Mexican line. He was formerly a butcher in Tombstone, and Judge Alex. Campbell states that he was known there as a buyer of stolen cattle. "He is one of the worst scoundrels in the Territory," says the Judge. Others vouch for him as a notorious fence. Geronimo and band first consulted Gen. Crook on the 25th. On the 26th Tribolett began supplying them with liquors—not, it is believed, from the ranch, but from a supply hidden near the Apache stronghold. On the 27th all the Apaches surrendered, and some were seen to be drunk. After the surrender, General Crook posted back to Fort Bowie, and Maus started with the prisoners. There seems no doubt that all were sincere in surrendering. They were sick of living like coyotes. They came along all right to the north of the line, where the Tribolett ranch is. He smuggled more whisky into them and played on their fears, doubtless suggesting that they would be hung. It is regretted that Tribolett could not be dealt with, as the chief trouble in settling the Apache question comes from him and his sort.

It only requires presence on the ground to see how awfully we have been imposed upon by the Associated Press dispatches from fevered newspaper offices in Tucson and Tombstone.

LUM.

Los Angeles Times, April 2, 1886:

SPECIAL DISPATCH TO THE TIMES.

FORT BOWIE, A.T., April 1.—Gov. Zulick and party returned to Tucson to-day. No demand was made on Crook, and the sole object of their visit was pleasure, and perhaps political clap-trap.

Lieut. Faison and the captives will arrive to-morrow.

LUM.

[16]Faison (1860–1940), was born in North Carolina. A West Point graduate, he became a Brigadier General, saw much service against Filipino insurrectionists, and in World War I took part in the Ypres-Lys and Somme offensives, earning the French Legion of Honor.—George W. Cullum, *Biographical Register of the Officers and Graduates of the U.S. Military Academy at West Point, N.Y.,* 8 vols. (Boston, Houghton, Mifflin and Co., 1891–1950), hereafter: Cullum; Francis B. Heitman, *Historical Register and Dictionary of the United States Army, from its Organization September 29, 1789, to March 2, 1903,* 2 vols. (Washington, Government Printing Office, 1903), hereafter: Heitman; *Who Was Who in America,* Vol. I, 1897–1942 (Chicago Marquis Who's Who Inc., 1966).

[17]There were perhaps eight Tribolets living in the area at this time, mainly around Tombstone. It is not certain that they were all brothers although all immigrated to this country from Switzerland; some may have been cousins of the others. It also is uncertain which Tribolet figured in the famous incident attending Geronimo's last bolt, but it probably was Robert (1861–1895). The Tribolets followed various occupations.—Information from Patience T. Wilson.

Charles F. Lummis and friend, Theodore Roosevelt.

Charles F. Lummis, probably about the time he went to Arizona.

Charles F. Lummis at his desk.

CHAPTER II

Los Angeles Times, April 6, 1886:

AT FORT BOWIE

The Apache Matter as Seen on the Ground

GERONIMO'S SURRENDER AND ESCAPE.

*A Sharp Set-Back on the Heels of a Glorious
Success—The Silent, Grim Old Soldier.*

[STAFF CORRESPONDENCE OF THE TIMES.]

FORT BOWIE, A.T., April 1, 1886.—I passed Tuesday night very handily at Bowie station with Lieut. McDonald, the herculean and intelligent commanding officer of the cavalry stationed there. He came from Fort Yuma last December, when so many troops were called from the coast at Gov. Zulick's outcry. Gen. Crook never sent for them and never wanted them. He had already as many men as he could use. I find among them all, however, that intelligent appreciation of the Apache problem which is almost non est among straight civilians. Lieut. McDonald has seen Indian service himself, and gave me many good pointers. In the morning, after a very creditable cavalry drill, he had a good horse brought up by a mounted orderly. My blankets and papers were strapped to the cantle, and I set out across the plain with the orderly, headed for Fort Bowie, which nestles in the mountains fourteen miles southeast of the station. The road is a good one, and we cantered along easily through the hot forenoon, troubled by nothing but a bronchial affliction—for water. At noon we had surmounted the three miles of tame cañon, and rose into

THE POST,

which lies on a rather sharply-sloping bench of the mountain side, while down through a gap in the hills one looks across the weird plain to a purple range 53 miles away. The post is called Fort Bowie, though there is no fort and no fortification. It is hemmed in by ranches on every side, and stands at an altitude of 4781 feet. Behind it is the inevitable crag—in this case "Ellen's Dome" [Helen's Dome]—from which a maiden threw herself to escape from the Indians. You know the mountainous country which cannot boast some such legendary cliff, is poor indeed. Around the generous plaza stand big, substantial adobes; and at the farther corner from the entrance, a French-roofed frame building of some pretentions, the residence of [brevet] Col. [Eugene B.] Beaumont,[1] Commander of the Post. Good water is pumped by steam-power

---

[1] A West Pointer from Wilkes-Barre, Pennsylvania, Beaumont (1837–1926) won a Medal of Honor in the Civil War and had long service with the 4th Cavalry. He was of distinguished

from an adjacent hillside. There are about 100 soldiers stationed here, and I fancy they have a very fair sort of a time. The discipline is Number One, and everything is trim and taut as a man o'war.

We rode up to headquarters at once, and were met by Maj. C.S. Roberts,[2] Adjutant General. He served on Crook's staff during the war, and is as genial a gentleman as one would care to meet. Gen. Crook and Gov. Zulick were having a pow-wow in Col. Beaumont's house, and did not conclude till one o'clock. Then

THE GREAT INDIAN FIGHTER

emerged and strode down to the office where we were waiting. He read my credentials, welcomed me briefly, spoke a few non-committal words in answer to leading questions, and said: "For any information you may wish, I will turn you over to Capt. [John G.] B[o]urke.[3] He is the best able to talk to you, and can give you everything accurately."

I like the grim old General. There is that in him which makes one want to take off one's hat. There never was a soldier who fought against heavier odds with a stiffer upper lip. He has the same patient, persistent, uncomplaining and unapologetic doggedness that was Grant's fundamental characteristic. To-day the most prominent figure in the army—the only one in the field—he occupies a larger place in public discussion than any other General. And in this exposed position, one of the fiercest fires is centered on him that ever whistled about a soldier's ears. Since the war none of its prominent commanders has been more persistently, more savagely, more cruelly hounded by jealousy, opposition and many another masked influence, than has Crook. Almost without exception the Territorial papers have damned him—not with "faint praise," but with bitterest invective. He has been cursed at, belittled and lied about, his policy misrepresented, his acts distorted, and alleged acts of his

---

ancestry, companionable, a good guitar player, composed his own ballads, was a fine mimic and excellent story teller. He and Lummis would have had much in common.—Robert G. Carter, On the Border With Mackenzie . . . , (New York, Antiquarian Press, Ltd., 1961).

[2]Cyrus Swan Roberts (1841–1917), born in Connecticut, had a good Civil War record and had accompanied Crook to his conference with Geronimo in late March, taking along his 13-year-old son, Charlie, who was later to become a Brigadier General himself. Charlie kept an interesting diary of the meeting, extracts from which are in the Arizona Historical Society library. Both Cyrus Roberts and his son are pictured in the famous photograph of that conference.

[3]John G. Bourke (1846–1896) needs little introduction to those interested in southwestern military history; his *On the Border with Crook* (New York, Charles Scribner's Sons, 1891), is the cornerstone for any study of the subject. In addition to being Crook's Boswell, Bourke was an ethnologist of respectable accomplishment, a gallant officer, a raconteur and skilled historical writer. He died too soon.

made up out of whole cloth. Some of these lies have already been nailed—as the one about his surrendering to the Apaches in the Sierra Madre in 1878 [*sic*].[4] Others have not been; and some never will be. He is a soldier, not a war correspondent. Let the lying go on as it will, telegraphed from end to end of the country—but he never opens his mouth. He is here to fight, not to justify himself.

IT IS LIKE PULLING TEETH

to get anything out of him, and in his own defense he will not let out a word. He feels it, no doubt—he would be more than man if the poisoned shafts did not sting. Yes, the old man carries other wounds than the Apache arrow that still rankles in his thigh—but you would never know it from him. Every time I look at him some half-forgotten lines come to mind. They describe the old grey wolf at bay. Who wrote them, you, who are near Bartlett's "Familiar Quotations," may hunt up:

> "And when the pack, loud-baying,
> His bloody lair surrounds,
> He dies in silence—biting hard
> Amidst the yelping hounds."

Not that there is anything of the wolf about Crook so far as ferocity is concerned. He is not even bluff, but as kindly as he is reticent. But it is the same unwhimpering grit, the same deathless hold. The General is a tall, well-knit man, without an ounce of superfluous flesh. He is straight, but does not convey that impression, for his well-turned head has that peculiar droop of the habitual hard-thinker. It is as though the weight of care and thought behind the seamed forehead dragged it forward from its poise. The deep, clean lines that mark his face are further tokens of the hard brain-work he has put into this campaign. He is an indefatigable worker, and keeps at the knotty problems all day and well into the night. His heavy brown beard is again usurping his chin, which a few weeks ago was shaven. His forehead is high and broad; his eyes clear and penetrating; his nose large and a very strongly hooped aquiline. He wears nothing to denote his rank or even his profession, but paces thoughtfully up and down the porch in a plain, but neatly-kept civilian suit, topped by a big, buff slouch hat. At Geronimo's surrender, the other day, the General wore a duck coat and common canvas pants. I dined with him yesterday and though he was inclined to respond in his laconic way to general

---

[4]This persistent rumor concerning Crook's great Sierra Madre expedition of 1883 is discussed in detail in Thrapp, *General Crook and the Sierra Madre Adventure.* It probably had some faint basis in fact, but even if partly true, was to the distinct credit of the General, rather than the reverse.

conversation, he was not to be pumped on the matter in hand. He didn't appear to evade, either, but simply said there was "nothing to tell."

THE STATE OF THE CAMPAIGN.

When I wrote you night before last, we all supposed the campaign at an end. The hostiles had surrendered unconditionally to the "Grey Fox" (as the Apaches call Crook), and were on their way to the fort as prisoners. It was only upon arriving here yesterday noon that I learned the unwelcome news of Geronimo's escape. The particulars are rather meager, but we will know more to-morrow, when Lieut. Faison will get in with the rest of the captives. I have now the full particulars of the surrender, also, and will communicate them as fast as may be, with the speeches made on that occasion by the Apaches, taken down verbatim on the spot.

GOING TO THE CONFERENCE.

Having received the news from Lieut. Maus that Geronimo desired to surrender, Gen. Crook immediately started, on the morning of March 22nd, for the San Bernardino rancho. This lies partly in Arizona, but mostly in Mexico, near the point where the lines of New Mexico, Old Mexico and Arizona come together. He was accompanied by Capt. B[o]urke, Major Roberts and Master Charlie Roberts, a bright and nervy lad of eleven [*sic*]. They traveled in an ambulance, with relays and no escort, reaching the line in 90 miles. Thence they went by saddle 25 miles in Sonora, arriving at Lieut. Maus's (Maws's) camp on the morning of the 25th. Geronimo and his band were camped in an impregnable stronghold upon a cañon-cleft hill a quarter of a mile from Maus, and saw them come. He sent word that he wanted to come down and have a talk with Crook. Before a reply could be got to him, he came down, accompanied by Nachita (commonly called Nachez)[5] and several bucks. Geronimo said he wanted to talk with Grey Fox. Crook replied, "I am not here to talk with you, but to hear you talk." Geronimo began a long-winded palaver, explaining why he had gone on the war-path. Crook promptly shut him up, saying, "Geronimo,

YOU ARE A LIAR.

You have lied to me once, and I cannot believe you any more. You must decide

---

[5]Nachez, whose correct name probably was Naiche, was born about 1856 and died in 1921, the last and youngest of the Chiricahua chiefs, who outlived all the others. He was the youngest son of Cochise and like his father a hereditary chief; Nachez descended also from Mangas Coloradas. Geronimo, who was a war leader, not a chief, deferred to him, although Nachez never was as positive a figure or as belligerent a warrior as Geronimo.—Gillett Griswold, *The Fort Sill Apaches: Their Vital Statistics, Tribal Origins, Antecedents* (unpublished manuscript at the Fort Sill, Okla., Museum), hereafter: Griswold.

at once to surrender unconditionally, or to stay in the mountains. If you decide to stay in the mountains, you may be sure that every last man of you will be hunted down and killed, if it takes us fifty years. Go back to your village and think it over, and let me know what you decide. I want no delay about it."

Geronimo was terribly taken aback at this cold reception to his overtures. He pulled nervously at a string, and big drops of sweat stood out on his hands and forehead. He went back with his companions to their fastness to chew the cud of deliberation. During this conference, Chihuahua,[6] the brains of the band, got back with his bucks from a raid in Sonora, driving a few ponies, and joined Geronimo on the hills.

CHIHUAHUA WEAKENS.

On the evening of the 26th, Chihuahua sent word to Gen. Crook that the bucks had been talking together all day, but had not yet reached a decision. But it made no difference what the rest did—*he* was going to surrender anyhow, and would come in that night, with his followers, if Crook would allow. But the Grey Fox was too shrewd for that. There was the very sort of leaven he wished to be working in the Apache camp; and he sent word to Chihuahua that none would be received until all surrendered.

THE SURRENDER.

And next day they all came in—that is, all the chiefs, Geronimo, Kut-le,[7] Chihuahua, Nana,[8] and Nachita, with a few bucks—and met Crook. It was an interesting scene, down amid the savage wildness of the Cañon de los Embudos. Besides Crook there were present Capt. B[o]urke, his right-hand man;

[6]Born about 1822, Chihuahua died in 1901. He was a full brother of Josanie (Ulzanna), and closely related to other prominent warriors. He rose to prominence through intelligence, courage, and ability and figured in such noted escapades as the Loco extraction from San Carlos in 1882 and the Chatto raid of 1883. Chihuahua was one of the most famous of the Chiricahuas. He died at Fort Sill.—Thrapp, *Dictionary of Frontier Characters;* Griswold.

[7]Kut-le (Kat-le; Kuthli), considered by soldiers of 1886 to be a chief, was not widely mentioned in reports of Indian hostilities, and Apache descendants of the Chiricahuas—Mimbres today do not recall his name. Perhaps he was the individual listed by Griswold as (Benjamin) Colle, or Cah-leh, who served in Company I, 12th Infantry during the Alabama exile and was killed June 23, 1894, near Mt. Vernon Barracks. His approximate year of birth is not given.

[8]Nana was one of the better-known Apache war leaders because he continued his belligerent activities into a most advanced age. Born about 1800, he died at Fort Sill in 1896. Nana was a Mimbres Apache, a close associate of Victorio, and later, after the melding of the remaining Mimbres with the Chiricahua, sometimes accompanied Geronimo. One of his wives was Geronimo's full sister. In 1881 Nana undertook a legendary raid across southern New Mexico, which Lummis later describes. Nana was a most astute and able leader.—Thrapp, *Dictionary of Frontier Characters;* Griswold; Harold Miller, "Nana's Raid of 1881," *Password,* Vol. XIX, No. 2 (Summer, 1974), 51–70.

Major Roberts and son, Lieut. Maus, Lieut, Faison, Dr. [Thomas B.] Davis,[9] ex-Mayor [Charles M.] Strauss of Tombstone, Mr. [Thomas] Moore,[10] master of transportation; Mr. [Henry W.] Daly,[11] and about eighty of Maus's Apache scouts—not to mention Mr. [Camillus S.] Fly,[12] a nervy photographer from Tombstone, who had gone into Geronimo's fastness the day before (prior to the surrender), and "took" the whole place, and everyone in it. There were also the interpreters, José María, Concepción, Antonio Besias and Mr. Montoya.[13]

CHIHUAHUA SPOKE FIRST.

A literal translation of his speech follows. It almost reminds one of a paragraph from Cooper. The Apache has a blarney stone of his own, and when

[9]Davis, born in 1844 in Indiana, was a brother of a noted Civil War General, as Lummis later points out. He had been an assistant surgeon with the army since 1869, served with Mackenzie in West Texas, served on Sioux operations in the north, and reached Arizona in 1882. He accompanied the Crawford expedition into the Sierra Madre, caring for Crawford, Tom Horn, and Mexican casualties of the ensuing fight. Later, as surgeon at the San Carlos Reservation, he attended Al Sieber after he was wounded in the Apache Kid outbreak, and an occasional victim from the Pleasant Valley War. After resigning from the army, he established a practice at Prescott, Arizona.—Thrapp, *Dictionary of Frontier Characters.*

[10]Crook thought Tom Moore (1832–1896) the greatest mule pack expert of his time, and perhaps he was correct. Crook always paid special attention to his pack animals and depended upon Moore in every western theatre in which he operated to keep them in top shape. Moore was a brother of saloon-buster Carrie Nation and is described in some detail later on by Lummis.— Thrapp, *Dictionary of Frontier Characters;* Agnes Wright Spring, "Prince of Packers," *True West,* Vol. XVIII, No. 1 (Sept.-Oct., 1970), 24–25, 42–46.

[11]Next to Tom Moore, Daly (1850–1931) was the most famous Army packmaster of the latter-day Indian wars, his service continuing through World War I. Daly was born in Ireland. He wrote manuals on pack transportation for the army and articles [on the Indian wars] of sometimes dubious reliability for popular magazines. He died at San Diego, California— Thrapp, *Dictionary of Frontier Characters.*

[12]Fly (1850–1901) was made famous by his photographs of the wild Chiricahuas taken on this adventure. His Tombstone studio was adjacent to the site of the Earp-Clanton OK Corral confrontation. In later years Fly served as Cochise County, Arizona, sheriff.—Thrapp, *Dictionary of Frontier Characters.*

[13]Much confusion exists about the interpreters, though they were key links in any meeting of Indian, Mexican, and white minds; mistranslation of sometimes obscure concepts from one tongue to the other, even of single words, might result in disaster, and sometimes did. "José María" may have been José María Montoya, perhaps not; Barrett said his last name was Yaskes. Fly's photograph of the Crook-Geronimo conference shows as interpreters: "Concepción, José María, Antonio Besias and José Montoya." Howard wrote that Concepción appeared to be about twenty-five in 1871, and was "a queer looking little man, half Mexican, half Indian." Other evidence indicates he was at Fort Sumner, New Mexico, in 1863 and later at Camp Goodwin, Arizona. Howard said his last name was Equierre, his father being Mexican. Bourke said that, like José María, Besias had been captured as a child and raised among the Apaches. Montoya was reported to have accompanied Lieutenant Charles B. Gatewood in the August interview with Geronimo which resulted in that Indian's final surrender.—O.O. Howard, *My Life and Experiences Among Our Hostile Indians* (Hartford, Conn., A.D. Worthington & Co., 1907); O.O. Howard, *Famous Indian Chiefs I Have Known* (New York, The Century Co., 1908);

he has a very big ax to grind, can flatter with the best of us. Chihuahua was talking for his life, which may excuse his poetical flights. He said:

"I am very glad to see you and have this talk with you. It is as you say—we shall always be in danger so long as we remain out here. But I hope from this on we may live better with our families, and not do any harm to anybody. I am anxious to behave. I think the sun is looking down upon me, and the earth listening. I am thinking better. It seems to me that I have seen the one who makes the rain and sends the winds—or he must have sent you to this place. I surrender myself to you because I believe in you, and you have never lied to us. You do not deceive us. You must be our God. I am satisfied with all that you do. You are, I think, the one who makes the green pastures, who sends the rain and commands the winds. You must be the one who sends the fresh fruits that come on the trees every year. There are many men in the world who are great chiefs and command many people; but you must be the greatest of all, or you would not come out here to see us. I want you to be a father to me and treat me as your son. I want you to have pity on me. There is no doubt that all you do is right, because all you do is just the same as if God did it. All you do is right. So I consider you to be. I trust in all you say. You do not lie. You do not deceive. All the things you tell us are so. Now I am in your hands. I place myself at your disposal. I surrender myself to you—do with me as you please. I shake your hand (grasping Crook's hand). I want to come right into your camp with my family, and stay there. I don't want to be away at a distance—I want to be right where you are. I have run in these mountains from water to water. I never found a place where I could see my father or my mother, till to-day, when I see you, my father. I surrender to you now, and I don't want any more bad feeling or bad talk. I am going over to stay with you in your camp. Whenever a man raises anything, even a dog, he thinks well of it and tries to raise it right and treats it well, even if it *is* a dog. So I want you to feel towards me and be good to me. Don't let people say bad things about me. Now I surrender to you and go with you. When we are traveling together on the road or anywhere else, I hope you will talk to me sometimes. I think a great deal of Alchesay and Kowten-nay[14] [two of Crook's scouts], and they think a great deal of me. I hope some

S.M. Barrett, ed., *Geronimo's Story of His Life* (New York, Duffield & Co., 1906); John G. Bourke, *On the Border.*

[14]Alchesay and Ka-ya-ten-nae (Kowtennay; Kowtenny) were Apaches upon whom Crook depended heavily to influence the wild Chiricahuas to come in. Alchesay was a White Mountain chief, "the most prominent and important" man of his people, and had won a Medal of Honor when serving as scout under Crook in the 1872–73 campaigns. He was born about 1853 and died in 1928.—H.B. Wharfield, *Alchesay* (El Cajon, Calif., pp. 1969). Ka-ya-ten-nae was a Mimbres Apache, born about 1861, who some said succeeded to chieftainship of his people after the death of Victorio in 1880. He surrendered to Crook in the Sierra Madre in 1883, but reportedly threatened Second Lieutenant Britton Davis and was sent to Alcatraz Island, California, then a military prison. Crook retrieved him to help bring in the hostile Chiricahuas.—Griswold; Thrapp, *The Conquest of Apacheria.*

day to be all the same as their brother, and that you will think the same of me as you do of them. I would like you to send my family with me wherever you send me. I have a daughter at Camp Apache [the reservation] and other relatives of my companions and me. Wherever you want to send me, I wish you would also send them."

I have remarked that Chihuahua is the brains of the band. You will see from the above that he is also the orator, and no slough at "giving taffy." Gen. Crook told him that his family should accompany him if they wished.

THEN NATCHITA TALKED,

in words which are, by interpretation, as follows: "What Chihuahua says, I say. I surrender to you, just as he did. I surrender to you just the same as he did. What he has said, I say. I give you my word, I give you my body. I surrender. I have no more to say than that. When I was free, I gave orders, but now I surrender to you. I throw myself at your feet. You order now, and I obey. What you tell me to do, I must do."

GERONIMO FOLLOWED:

"Two or three words are enough. I have little to say. I surrender myself to you (shaking Crook's hand). We are all companions, all one family, all one band. What the others say, I say also. Now I give myself up to you. Do with me what you please. I surrender. Once I moved about like the wind. Now I surrender to you, and that is all. I surrender to you, and want to be the same as if I was in your pocket. Now I feel like your brother, and Kowtenny is my brother also. I was very far from here. Almost nobody could get to that place. But I sent you word. I wanted to come in here, and here I am. Whatever you tell us is true. We are all sure of that. I hope the day will come when my word will be as strong with you as yours is with me."

Nana and Kut-le then surrendered. Crook asked the chiefs if the surrender included all their people. They said it did. Then followed a general confabulation, in which the prisoners asked to be taken back to the fort by easy marches, as their stock was worn out. Crook promised. The bucks, squaws and children came in during the night, and before morning all the 92 hostiles were prisoners. Gen. Crook then hastened back here, leaving Lieut. Maus and his command to bring up the prisoners.

In my next you shall hear of the sensational escape of Geronimo, Nachita, 19 other bucks and 13 squaws, and the causes.

Gov. Zulick and his crowd, now sober, returned to Tucson this morning. Gen. Crook says there was nothing official in their visit—just a little *pasear* out of curiosity. He has nothing to say as to what he will do with his prisoners, whom I hope to interview to-morrow. *A más ver.*

LUM.

Los Angeles Times, April 3, 1886:

THE APACHE CAPTIVES

Chihuahua says it was all Geronimo's Fault.

[SPECIAL DISPATCH TO THE TIMES].

FORT BOWIE, A.T., April 2.—At noon to-day Lieut. Faison arrived with fifty-eight prisoners, including Chihuahua, Josanie[15] and Katle, three of the worst leaders of the hostiles; also old Nana and a white boy, a captive. The Apaches seemed very glad to come in. They are camped with Gen. Crook's Indian scouts, half a mile west of the fort.

This afternoon Chihuahua had a conference with the General, lasting an hour. Chihuahua said that he was very happy to get back to his wife and children, and wanted always to live quietly with them. He knew he had committed many depredations, but Geronimo was to blame for all. Geronimo had dragged them off the reservation by lies. He thought that Geronimo would never come in now. He said: "I've thrown away my arms. I'm not afraid; got to die sometime. If you punish me very hard it's all right, but I think much of my family. You and almost all your officers have families, and think much of them, so I hope you will pity me and will not punish too hard."

Gen. Crook told him to go back to camp, get rested and they would have another talk in a day or two. Gen. Crook says there is much truth in what Chihuahua says about Geronimo getting them off the reservation.

LUM.

ARMY CHANGES

General Crook Removed From the Department of Arizona

WASHINGTON, April 2.—A General Order was issued by the War Department this afternoon. . . .

Brigadier-General Crook was to-day relieved from the command of the Department of Arizona, and assigned to the Department of the Platte, formerly commanded by [Major] General [Oliver Otis] Howard [now assigned to the Division of the Pacific]. Brigadier-General N.A. Miles, now in com-

---

[15]Josanie (or Ulzanna; Alzanna) was born about 1821 and died in 1909. He was one of the best known—to the whites—of Chiricahua warriors because of the slashing raid he led in November and December, 1885, into southern Arizona and New Mexico. A brother of Chihuahua, he died at Fort Sill.—Griswold.

mand of the Department of the Missouri, has been assigned to the Department of Arizona.

———

Los Angeles Times, April 4, 1886:

FROM FORT BOWIE

Lieut. Maus Compelled to Abandon the Pursuit of Geronimo.

[SPECIAL DISPATCH TO THE TIMES].

FORT BOWIE, A.T., April 3.—Lieut. Maus and scouts arrived this afternoon, bringing in two more bucks of Geronimo's band, one a brother of Chipuesa[16] and the other a brother of Kowtennay.[17] They had followed Maus and surrendered voluntarily. He thinks that more hostiles are likely to do the same. He had to abandon the trail of Geronimo, after following it sixty miles, to near Fronteras, Sonora.

After leaving his camp on the night of the 29th, the hostiles hurried through the most impassable mountains, stabbing their only horse after a short distance and breaking their trail on rocks. In all the sixty miles they did not once camp. Near Fronteras they scattered in every direction, making for their old strongholds in the Sierra Madre. Maus had but four days rations. His men and stock were worn out, and he had to drop the trail.

Probably there will be no more operations until Gen. Miles arrives.

LUM.

Los Angeles Times, April 6, 1886:

ON THE TRAIL

Another Party in Pursuit of Geronimo

[SPECIAL DISPATCH TO THE TIMES].

FORT BOWIE, A.T., April 5.—Frank Bennett,[18] a scout, arrived last night from Lang's Ranch.[19] He left there Thursday. Capt. [Joseph Haddox]

---

[16]This man is unidentifiable today.

[17]Probably Kinzhuna (1866–1941), who was exiled with the other Chiricahua-Mimbres in 1886. He was related by marriage to Mangas Coloradas and Victorio. Kinzhuna attended Carlisle Indian School in 1887–89.

[18]See Lummis's special article on Bennett below.

[19]Lang's Ranch, in the extreme southwest corner of New Mexico on the Old Mexico line, was established by Billy Lang, who was killed with Newman Clanton and three others in August of 1881 in Guadalupe Canyon, southwestern New Mexico, by a Mexican "posse" of sorts. The

Dorst,[20] having just received news of Geronimo's escape, was starting with 135 men to follow the hostiles into Mexico. He took thirty days' rations and said he would go on for twenty days at least. The six months' enlistment of his Indian scouts expires this month, so that he cannot make a longer campaign.

Santiago McKinn, the 11-year-old white boy, the Apaches' prisoner taken with Geronimo's band, will be sent home to-morrow. It is learned that his parents were not killed, but reside at Hot Springs, at Hunter's, N.M., near the railroad from Deming to Silver City. During his half-year of captivity the lad had grown fully Indianized. He joins their sports, and will have nothing to do with the whites. He understands English and Spanish, but can hardly be induced to speak in either. He has learned the Apache language and talks it exclusively.

Sheriff [Robert] Hatch, of Cochise County, came up the other day, with a warrant for Geronimo and forty-one John Does. General Crook told him to make his demand in writing. He did not do so till he got back to Tombstone. To-day a written demand was received from him for Chihuahua and the other renegades. General Crook replied that the Chiricahuas are held as prisoners of war, under instructions from Washington, and will not be given up.

LUM.

―――――――

Los Angeles Times, April 7, 1886:

THE APACHES

The Prisoners to be Sent to Fort Marion, Fla.

[SPECIAL DISPATCH TO THE TIMES].

FORT BOWIE, A.T., April 6.—The Chiricahua Apaches who surrendered to Gen. Crook March 29, and arrived here April 3, in charge of Lieut. Maus, will be sent to Fort Marion, St. Augustine, Florida, as prisoners of war. The squaws and children go too, making seventy-six in all. Company E, Eighth Infantry,

ranch continued to be identified by its former owner's name, however, and at this time was a familiar stopping place because there was water there.—Thrapp, *Dictionary of Frontier Characters.*

[20]Dorst (1852–1916) was born at Louisville, Kentucky, and after graduation from West Point joined the 4th Cavalry in 1873. He served in West Texas, being Mackenzie's "favorite officer," it was said. He came into disfavor with President Theodore Roosevelt for criticism of the Rough Riders, and Roosevelt "never would make Dorst a General Officer," a promotion he deserved. Dorst saw much service in the Philippines against insurrectionists. He retired in 1911.—*Who Was Who;* Heitman; Cullum; Joseph Dorst Patch, *Reminiscences of Fort Huachuca, Arizona* (Washington, D.C., pp., 1962), hereafter: Patch.

go as a guard, under command of [First] Lieut. J.R. Richards,[21] Fourth Cavalry. The prisoners are giving a grand dance to-night at their camp, as a farewell to Gen. Crook's Chiricahua scouts. The prisoners know that they are going away, but don't know where. They take it very philosophically. Chihuahua told Gen. Crook to-day that wherever he sent him it was all right, but he hoped he would not suffer imprisonment too long, because he would lose a wagon he had at the reservation.

Santiago McKinn, their eleven-year-old white captive, was sent home to-day. He would not leave the camp with a white man, and had to be brought into the fort by Chiricahuas. He bawled loudly when told that he was to be taken back to his parents, and said he always wanted to stay with the Indians.

LUM.

Los Angeles Times, April 8, 1886:

The Apache prisoners, in charge of government troops, have been started from Fort Bowie to Fort Marion, Florida. That poor Florida should be [afflicted] in the same year with a cold snap and a batch of Apaches is too much!

Los Angeles Times, April 8, 1886:

THE APACHE PRISONERS.

They Start for Florida—The Results of Crook's Campaign.

[SPECIAL DISPATCH TO THE TIMES].

FORT BOWIE, A.T., April 7.—Chiefs Chihuahua, Nana, Kutle and Alzanna [Josanie], with seventy-two other Chiricahua-Apache prisoners, bucks, women and children, left here at 11:30 this morning for Fort Marion, Florida, as prisoners of war. Gen. Crook went to Bowie station with them on a buckboard. Twenty-five Apache scouts escorted them. All went cheerfully, though understanding what is to be done with them. Chihuahua was riding around all the morning hurrying them in preparations to move.

Two weeks ago ninety-two hostiles, including six chiefs, were on the war-path. Now there are but thirty-four, of whom fourteen are squaws and two chiefs. This putting their families and allies far beyond reach will be a severe blow to them. Geronimo's wife and daughter and some of Natchez's children are among the prisoners. Sixteen of the seventy-six were captured by Lieut. Davis[22] some months ago. To have run down and killed or captured the other fifty-eight in the ordinary course of Apache warfare would have cost many American lives, and twenty-five to fifty thousand dollars per head. Crook has therefore accomplished something.

LUM.

OMAHA, Neb., April 7.—Capt. J.G. Bourke of Gen. Crook's staff, who accompanied the remains of Capt. Emmet Crawford from New Mexico to Kearney, Neb., arrived in Omaha today to perfect arrangements for Crawford's funeral, which will take place at Kearney, on Sunday. . . .

[21]James Russell Richards Jr. (1854–1914) was born in Virginia, graduated from West Point, served in the 9th Cavalry briefly in 1878, then transferred to the 4th Cavalry from which he retired a captain in 1896.—Cullum; Heitman.

[22]Lummis here means Captain Wirt Davis, 4th Cavalry. Davis (1839–1914) was born at Richmond, fought on the Union side in the Civil War, took part in a number of Indian operations thereafter, winning several brevets. He was considered "the best pistol and carbine shot in the Army" and retired in 1904 as a Brigadier General.—Thrapp, *Dictionary of Frontier Characters*.

Geronimo, the symbol of Apache resistance to white encroachment in the Southwest. Photographed by Camillus S. Fly of Tombstone in 1886.

Geronimo and Natches (Naiche) at Fort Bowie in late summer of 1886 after their surrender. Photograph from the Sacks Collection, Arizona Historical Foundation.

Fort Bowie in 1886.

Ruins of Fort Bowie as they look today. It is now a National Historic Site.

Colonel Eugene Beaumont, commander of Fort Bowie when Lummis was there in 1886. Photograph courtesy the Arizona Historical Society.

John G. Bourke, Crook's dependable aide with whom Lummis often spoke and from whom he obtained much of his background information on Crook. Photograph courtesy the Arizona Historical Society.

Commanding Officer's quarters at Fort Bowie when Lummis was there. Photograph from the Sacks Collection, Arizona Historical Foundation.

CHAPTER III

Los Angeles Times, April 9, 1886:

THE APACHES

What Makes Them Hard to Conquer.

A TOUGH COUNTRY TO CAMPAIGN IN.

*The most Savage and Elusive Warriors Alive—With Passing
Reference to the Premier Prevaricator.*

[STAFF CORRESPONDENCE OF THE TIMES.]

FORT BOWIE, A.T., April 5, 1886.—The California liar has amassed a monumental notoriety not honestly his own. He is *not* the Premier Prevaricator of the Universe. He does well for his gifts, but he's over-matched. The boss, unapproachable and supreme twister of Truth's caudal appendage isn't he of the g.c., but the fiery, untamed, mouthful Arizonian—the multitudinous gentleman who has been feeding the Associated Press with reports of the Apache campaigns, particularly the present one. Of these reports I believe it moderate to say that not one in fifteen has been approximately true. Most of this economizer of truth dwells in Tombstone—and by what scratch did Tombstone ever carom on the frigid facts? Part of him hangs out in Tucson— that arid aggregation of toughness, adobe and spare time—where people have too much leisure to tell the truth. He has also some members at-large in other parts of the Territory. And when he unbuttons his mouth, it shall be a pity if you don't get some news.

But, in honest and sober fact, the extra-Territorial papers have been

ABOMINABLY ABUSED

in the matter of war news from this section. No newspaper, until the TIMES, has had a representative anywhere near the field. The Associated Press has had no agent within one hundred miles of any fighting; nor has it sent any person even to headquarters for news. No dispatches have been sent out from anywhere by any actually posted person, until within a very few days. It is not the policy of this department to fight in the newspapers. Crook is a soldier, not a war correspondent; and he has, perhaps, carried his grim dignity to an extreme in not making public the facts that would vindicate him.

HOW WE HAVE GOT "NEWS."

Mayhap 'tis an errant "cow-puncher" who lopes into Tombstone, fills his hide with intestine-corroder, and begins to shoot off his war news. He hasn't been within fifty miles of the field, but a little thing like that doesn't bother him. He can tell you more about it in a day than Crook ever dared to know. "Bar-keep" takes it all in and retails it to the next cuss-tomer; the n.c. pours it into the elastic receptivity of the Associated Press agent, and the A.P. a-toots it to a gaping world. Or, perchance, some discharged mule-persuader from the military pack-train turns himself loose on the first unprotected settlement he strikes; and the rest of the story is carried out as per programme above. Yet, again, it is the gentle tin-horn gambler who cajoles his hours of ease by putting up a cold deck upon unsophisticated Truth, and deals a pat hand to the agent of the great news dispenser. These are not guesses, but plain statements of the way in which the "news" for which we pay has been born. I have in my hand at this moment specific clippings from just such sources and none other. But they were whooped up, all over the country as the latest news. I don't argue that it is absolutely impossible for drunken cowboy, deposed burro-beater or tin-cornucopia professor to some time stumble upon the truth. Accidents will happen. Nor do I know that the Associated Press is to blame. It has been imposed upon, like all the rest of us. Aside, however, from the merely *unreliable* sources from which the news has been drawn, there is a big, strong anti-Crook ring, of unknown periphery, of many diverse materials, of great evident weight, and homologous only in the desire to "down Crook." When I get time I'll measure the diameter of this ring as closely as may be; meantime, to a few general facts, which are all-important preliminaries to any statement as to this campaign.

No man can grip the full breadth of the situation, who does not, to start with, know this Arizona country, root and branch; and none can get even a finger in Truth's pie who has not a fair realization of the following physical facts! No campaign in the civil war, or in any of the northern Indian wars, was ever so entangled and crippled by topographical cussedness.

In the first place, then,

APACHEDOM IS A DESERT,

partially redeemable, it is true, by the future development of artesian and canal irrigation, and already dotted with semi-occasional oases. But I can lead you 500 miles, in a not palpably circuitous route, and in all that hideous stretch you shall see not one drop of water, save the precious fluid in our water-kegs. This arid desert is not one vast sea of drifting sand. It is one of the most mountainous sections of the whole country. And you will find square miles of it carpeted with the Etruscan gold of fragile poppies, and other miles of many another flower. The gray sage brush; the greasewood's glaucous green; the emerald daggers of the amole; the duller-hued bayonets of the aloe, topped with a banner of snowy bloom—these diversify all its valleys, while here and there loom the vast candelabras of the giant-cactus. All this is aesthetic but not filling. Ride twenty miles across your flowery plain, and you would swap your tongue for a sun-baked sponge. Ride fifty, without water, and you will do well, indeed, if you ever see sanity again. It is a country from which, *sans* water, a bunk in sheol would be a positive relief.

RAGGED RANGES.

Prime features of this section are the mountains. Fancy an irregularly, undulating but regularly thirsty plain of 300 miles, broken by but three or four subterranean water-courses. Upon this vast area [is] a wilderness of countless peaks and ridges, planted hap-hazard. Hunt the world over and you will find no more inhospitable and savage mountains. Shaggy, with sharp rocks and sharper cactus, they rise 500 to as many thousand feet above the circumfluent plain, their highest peaks wooded and snow-bound. You shall pass within a mile of such a hill, and have no more notion of its inaccessibility than a cow has of a hereafter. Try to climb the smallest and you will find out. There never have been but two animals which have loomed up as successes in scaling these rocks—the mountain sheep and the Apache. Either skips over them like rolling off a log. These ranges form such an "underground railway" as cannot be beat. A man of ordinary secretiveness could slink from Colorado to Mexico along these Apache trails and never be seen by human eye. None but the high-circling buzzard and prowling coyote would note his passing. Skulking through the mountains by day, dashing across the interjected valleys by night, he could be as unobserved as if he

BURROWED UNDERGROUND.

And even should some casual hostile glance detect him, he has but to shrink to yonder crest, and he is safe. He can kill 500 men as fast as they can come to him—himself almost absolutely unexposed. And then he can sneak back from sheltering rock to rock, until he is beyond pursuit. This is, of course, on the supposition that his foes are whites. He couldn't play that on the Apache. If a man who really hankered to hide out here got caught or killed, it would be because he was either a blamed fool or playing to very hard luck, and yet there are acres of good, rational people all over this country who fancy that all there is to this Apache business is to chase Lo over a field until he gets tired and then perforate him with a 45-70, or tie him up and bring him into camp in an express.

That's a [the] breed of geography they raise out here.[1] Now for

THE NATURE OF THE NATIVE.

Trying to hit the bull's-eye of this matter, at my limited verbal range, is like

---

[1] If Lummis's free-wheeling descriptions tended to exaggerate the geographical difficulties of the Arizona of his time, our more sophisticated generation tends to minimize them. We are insulated from harsh physical contact with the state's primeval elements by technological advances and today's denser population of the area.

trying to lasso a broncho steer with a yard of sewing-cotton. Hold the dictionary up by the tail, and still I can't shake out the vocabulary to phrase the facts. Language will scarce graze the skin—but here's a try.

The North American Indian, by-and-large, has never been notorious as a dude or an ass in war. Crude his methods may be, but they are effective. It has put the "superior" white man to his trumps to "get away with" him; and it never would have been done but for infinitely better weapons, later superior numbers, and a judicious use of whisky. Some tribes have naturally inclined to peace and endurance of wrong; some have fought fearfully at the pinch; and some are

BORN BUTCHERS—

hereditary slayers.

Foremost in the latter class has always stood the Apache. For warfare in his own domain he has been, and is to-day, without a peer. From time untold he has been a pirate by profession, a robber to whom blood was sweeter than booty—and both as dear as life. Untold generations before the Caucasian outpost encroached upon his Sahara, he was driving his quartz tipped shafts through agricultural Aztec, peaceful Pueblo or plodding *paisano.* The warlike tribes to his east and north, too, were represented in many a jetty lock at his belt. From Guaymas to Pueblo, and from San Antonio to where the Colorado laps the arid edge of California, he swept the country like a whirlwind. Of what he has done to keep his gory hand in since blonde scalps first amused his knife, I need not remind you now.

Not only is he the most war-loving of American Indians. He is also

THE BOSS WARRIOR.

He is strong to an endurance simply impossible in a more endurable country. He has the eye of a hawk, the stealth of a coyote, the courage of a tiger—and its mercilessness. He is the Bedouin of the New World. His horses will subsist on a blade of grass to the acre, and will travel 110 miles in twenty-four hours thereby, without dropping dead at the finish. He knows every foot of his savage realm better than you know your own parlor. He finds food and drink where we would perish for want of both. He has a fastness wherever you may strike him, and it is practically impregnable. Lay siege to him, and he quietly slips out by some cañon back door, and is away before you know it. The dangerousness of an Indian is in the inverse ratio of his food supply. His whole life an unceasing struggle to tear a living from nature, the Apache is whetted down to a ferocity of edge unattainable by the Indian of a section where wood, water and facile game are ready to his hand. Why, his six-year-old boy will ride a broncho farther in a day, and over rougher country, than you could ride the gentlest steed. These kids who were out with the hostiles were doing it right along.

But this is not all that puts the Apache at the head of his class—he has pals to

STAND IN WITH HIM.

From the outstretched arm of pursuit, he slides down into old Mexico as if the hills and valleys were a greased pole—but taking time to murder, rob and ravish in transit. He gets safely into Sonora; sells his stolen stock without any trouble; caches the stolen arms, ammunition and money; enjoys a genteel loaf in the Mexican Sierras until he is rested; swoops down upon hacienda and village, killing a few people and gathering up all the loot he can pack or drive; and flits back like a black shadow to his Arizona strongholds. The better class of Mexicans desire his extermination; even the lower classes sometimes organize against him; but he finds plenty of degraded natives to help him. The Mexican line is not only a line—it is a wholesale "fence." And safe to say, some poor, mescal-corned *paisano* is not the Apache's only pal. There are white Americans who batten up on his bloody booty. You will find them in Tombstone, Tucson, and many another place on either side of the line. If the source of the raid-causing whisky were published, there are some Arizona merchants who would writhe some—but at $20 a gallon they take their chances. It is to one of these beneficiaries of murder that the present lapse of a superb success is due.[2]

I have already told you about the sweeping and unconditional surrender by which the Gray Fox became possessed of Geronimo and his whole band, including every Indian off the reservation, except Mangus[3] and two or three bucks who have been out four years, and had nothing to do with the present campaign. Now for the unhappy sequel.

OUT AGAIN.

There is no doubt that Geronimo and his band surrendered in good faith. They had no other earthly reason for giving themselves up, but were tired of the war and glad to come in and take their chances. Whatever disposition

---

[2]Here again Lummis may have exaggerated, but not much. For a suggestion of the clandestine trade which kept the Apaches in "business" for many years, see Thrapp, *The Conquest of Apacheria*, 345–47, and Dan L. Thrapp, *Victorio and the Mimbres Apaches* (Norman, University of Oklahoma Press, 1974), 255–56. Of course the Apaches also were sinned against by being sometimes blamed for white thievery.

[3]Mangus was a son of the great Mangas Coloradas. He was born about 1846 and died at Fort Sill in 1901. He was seventeen when his father was killed, and with his people he gradually melded with the central and southern Chiricahuas, although he frequently remained aloof from them. Mangus came in with the others after the Crook expedition to the Sierra Madre in 1883, bolted with Geronimo in 1885, but did not surrender with him. In October, 1886, he brought his handful of followers out of Mexico northward to surrender at Fort Apache. He was intercepted and taken in by Captain Charles L. Cooper of the 10th Cavalry and sent to Florida.—Thrapp, *Dictionary of Frontier Characters.*

might be made of them, they knew that Crook would give them fair play. This absolute confidence of the Indians in his honor is almost as important a factor in Crook's success as his matchless knowledge of their traits. The hostiles would not have surrendered thus to any other man. All was serene; but one of the same

MALIGN INFLUENCES

which, from year to year, have fanned the savage spark to the blaze of war, again got in its work. The great San Bernardino rancho runs along the Sulphur Springs valley, from this side of the line, down many leagues into Sonora. Indeed, the surrender took place on it, twenty-five miles below our boundary. On this rancho, some 400 yards below the line, lives a Swiss-American, named Triboulett—long notorious in Tombstone as a "fence" for "rustlers." He was also tried, some years ago, for stealing barley from the government, at Fort Huachuca. It isn't easy, even yet, to convict a man in this Territory, and he got off. He is still deemed a "fence" and, infinitely worse, furnishes whisky to the Indians. He makes no secret of it, and snaps his fingers at protests. On the 26th, the day before the surrender, it was noticed that Geronimo and other bucks were getting pretty full. It has since been discovered that Triboulett had smuggled five five-gallon demijohns to a secret place near their fastness. Still, they surrendered all right, and came along handsomely as far as Smuggler's Springs,[4] where they camped on the night of the 29th. There they came in contact with

MORE OF TRIBOULETT'S WHISKY,

despite all possible precautions to keep them from it. Some of Maus's Indian scouts had smashed this white scoundrel's whisky barrels, and destroyed all the liquor in sight. That's a rather sarcastic commentary on our Nineteenth century civilization! Triboulett and his emissaries played also upon the fears of the prisoners, telling tbem they were putting their necks inside the halter. Savage as the Apache is, there are matters in which he is a perfect child. Take him in the night, especially when he is tipsy, and the veriest vagabond's ghost story will stampede him. Of course, you will understand, from what has already been said of the Apache, that Lieut. Maus's eighty-four men were entirely inadequate to surround, bind or disarm the ninety-two prisoners; and they were practically as free as ever. It would have taken 1000 men to make even a stagger at doing it, and even then, many a life would have been lost in the operation. At the faintest hint of either proposition, the Apaches would have been off like a flock of quail; and from the first cover their rifles would have sent back their defiance.

[4]Contrabandista Springs were three miles below the Mexican border near the San Bernardino River, which generally was dry at this season.

The conspirators succeeded; and that night, during a rainstorm, Geronimo and Nachita, accompanied by twenty other bucks and fourteen squaws—one an immature girl—

SLUNK OUT OF CAMP

noiselessly, and vamosed. They took their weapons, but only one horse. The prisoners had camped only a short distance from Maus, and no one knew of their departure until morning. If any martial reader of the TIMES thinks he could have held these drink-crazed demons there is a good chance for him now to come out here, drop a little salt on the fugitives, and end the war.

Governor Zulick and his crowd have gone home. I have the authority of the Governor and of General Crook for saying that no demand has been made to have any of the prisoners turned over to the civil authorities.

Am getting some mighty interesting notes about the killing of Captain Crawford[5] by Mexican troops last January. The facts have not been half published, and there is an apparent disposition somewhere not to have them. But they shall see type, if I never sell another fish.

LUM.

Los Angeles Times, April 11, 1886:

PRESIDENT DIAZ

His Message on the Killing of Capt. Crawford.

WASHINGTON, April 10.—Señor [Matías]Romero, the Mexican Minister, has just received the message the President of Mexico transmitted to the Mexican Congress on its convening on the 1st inst. A brief synopsis of what the President said of the encounter in which Capt. Crawford was killed was telegraphed from the City of Mexico. The following is a translation of that passage of the message:

On the 23d of January last the Governor of the State of Chihuahua, in a report by telegraph to the War Department, says that on the 11th said

---

[5]Emmet Crawford (1844–1886) was born at Philadelphia; his slaying by Mexican irregulars in January, 1886, was a celebrated incident of the Apache wars, and his demise was generally lamented. He enlisted in the Civil War, was commissioned in 1864, and eventually became a captain in the 3rd Cavalry. After brief service in Arizona, he was transferred north for operations against the Sioux, taking part in the Rosebud battle of June 17, 1876, and other affairs. Returned to Arizona in 1882, he was named military commander of the San Carlos Reservation. Crawford had a key role in Crook's 1883 Sierra Madre expedition. After the Geronimo outbreak he operated in Mexico against the Apaches.—Thrapp, *Dictionary of Frontier Characters.*

Mexican forces had had an encounter at a place called Tiopare,[6] in the mountains, with 200 Indians commanded by foreign officers, and had killed five persons, among them their captain (Crawford). On our side several persons were also killed and wounded. Among the former were the major commanding the troops, and a lieutenant. This incident was somewhat distorted by the North American newspapers, and this gave rise to the supposition in the United States that the killing of Captain Crawford in said encounter had been intentionally caused by our troops, as they (the papers) assert that they (the Mexican troops) fired against the officers of the United States Army knowingly. Public opinion was somewhat excited in both countries, and the two governments were compelled to order a careful investigation of the facts. For my part, and taking into consideration what has been ascertained up to the present time, I have the conviction that in said encounter our troops thought that they were fighting hostile Indians, because they were following the track of the savages and of cattle which had been stolen, and they could not possibly imagine that said Indians had been joined by others similar in aspect, and among whom were a very few soldiers or officers of the United States. The killing of the courageous and deserving officers and citizens of both countries is a very lamentable affair; but our troops, which were composed of citizens of the State of Chihuahua, will always have the excuse that they could not take as friends the Indians who were in front, when they well knew that, according to the agreement for the passing of troops over the frontier, only the regular troops of both Republics can pass reciprocally the boundary line when they are following the hot trail of hostile Indians.

Los Angeles Times, April 11, 1886:

"INJUNS!"

A "Times" Commissioner With Bucks and Bronchos.

ALSO SQUAWS, PAPOOSES AND PONIES.

An Inner View of Apache Life and Labor—The "Grey Fox" and the Work He Has Done.

[STAFF CORRESPONDENCE OF THE TIMES.]

FORT BOWIE, A.T., April 7, 1886.—Well, the agony is over for the time being. What new excitement we are to have can be told only by the last of the week and by General [Nelson A.] Miles. It is believed by some that the new

---

[6]Tiopare is on a tributary just west of the Aros (Haros) River, 26 miles east of Sahuaripa, Sonora.

commander will pick up the whole force at his disposal and fling it at Mexican space at once, in hope of hitting Geronimo with some scattering fragments. Others, looking at the circumstances more conservatively, opine that Miles will merely "lay for" the remaining renegades; and whenever he sees a head, hit it, but not indulge in preliminary beating of the bush. You see, the closing work of Gen. Crook's administration has put an entirely new face on the Apache situation. Instead of 110 raiders, as there were two weeks ago, there are now but 34. Instead of an army with six war chiefs, there is now a handful of men with two leaders—Geronimo the foxy talker, and Natchita, the hereditary chief, who is but a half-hearted warrior. Chihuahua, Nanay, Kut-le and Ulzanna (commonly corrupted to Hosanna [Josanie]) are now on the way, with their warriors, their squaws and their children, a band of 76 in all, to a Floridian prison. Instead of being hampered by babies, old women and numerous young squaws, the outstanding renegades have 14 able-bodied squaws to do their work, no children and no encumbrances. If it hung on the ragged edge of impossibility to catch them before, when so encumbered, what will it be now? But there is another matter. Having been pursued unceasingly for the last ten months, it is probable that they will now lurk in the far fastnesses of the Mexican mountains, and not trouble our side of the line for a considerable time. This is the more likely from the fact that all their people have been put beyond their reach. Had the captives been sent back to the Reservation, save a few of the ringleaders, the outstanding hostiles would probably have raided up there and got them out again, but now they have no show of getting reinforcements, nor even of recovering their families. This is apt to break them up. I am inclined to believe that if they could be communicated with to-day, and told the exact facts, they would start for Fort Bowie to-morrow—all but perhaps Geronimo and three or four of his intimate followers—to give themselves up. Unluckily, one might as well try to telephone to a comet.

At all events, that to-day's consummation scores a big

TALLY FOR GEN. CROOK,

every fair-minded person must admit. He has reduced the number of renegades on the warpath by four-fifths within a fortnight, without losing a drop of American blood. It is a creditable wind-up to a remarkable campaign, even though the Grey Fox is now removed from the power of absolutely winding up the matter by the final capture of the last hostile. In view of these facts, if any one wants to call this campaign a failure, of course there is nothing to hinder.

As one of the results of the farcical "news" that has been furnished throughout, we have heard nothing but Geronimo, Geronimo, Geronimo. One would fancy that old Jerry was the only Apache that has been off the reservation; and there is not much question but that it would have made a bigger impression on the public if instead of the seventy-six prisoners now rolling toward Fort Marion

GERONIMO ALONE

had been captured, and all the test of his band were still at large. The fact is, Geronimo is only one of seven chiefs who have been off the reservation with their families and followers. He is not even a No. 1 chief, but merely a war-chief, Nachita being the hereditary high muck-a-muck of the Chiricahuas. Nachita is an indecisive fellow, fonder of flirting than fighting, greatly addicted to squaws, and rather easily led by Geronimo, who is a talker from Jawville. Geronimo has not been the biggest fighter, the biggest schemer nor the bloodiest raider in the outfit at any time till now, when he has only the dude Nachita. Chihuahua is smarter, Nanay, Kut-le and Ulzanna more blood-thirsty and daring. Their bands have done more raiding and more mischief than Geronimo's. The only claim Geronimo has to his unearned pre-eminence of newspaper notoriety is that he is one of the originators of the outbreak. He is

NO GREATER AND NO WORSE

than several of his co-renegades.

When they arrived, with Lieut. Faison, the broncos (as the hostile Chiricahuas are termed, in contradistinction from the scouts), camped about three-quarters of a mile from the fort, behind a hill. They chose the bed of a dry arroyo, and settled down contentedly amid the tufts of bear-grass and cactus. Firewood and rations were hauled down to them, and they utilized both to the utmost. The women and children captured by Capt. Davis and Capt. Crawford, some months ago, and since held prisoners at this post, were let out, and rejoined their relatives and friends, with demonstrative welcomes. All but one, that is. There was one good-looking girl, who wept bitterly at the prospect of being released from the guard-house, and appealed to Gen. Crook to let her stay—a request which he granted. It seems that her husband was killed in one of the raids, last year; and she foresaw that if she rejoined the band, in which she had no relatives, she would be kicked from pillar to post, and have to work like a reporter, to be tolerated at all. The others, however, got along as sweetly as flies on a molasses can. They didn't do much, the first day, except to whoop up their fires and inter their rations. Next day, however, they began to change the aspect of things.

THE PATIENT SQUAWS

had knotted back the rambling bear-grass with strings of its own tough fiber, grubbed out the immediate cactus, dragged in branches of scrub oak and mesquite, and made semi-circular wind-breaks, four or five yards from point to point. Against these were laid the blankets, the solid canvas and other "furniture." Then here and there you would see a squaw bracing up some tall stalks of the century plant and lashing them together at the top, while her sister in drudgery was stitching together big lengths of unbleached muslin. A couple of days of this hard work—to which the bucks lent their moral support by industriously gambling at koon-kan for cartridges, money or ponies, or promenading along the fort—the camp was complete and a work of art. The Apache weekly *Scalping Knife* would doubtless have filled some columns with a description of this "growing young Chiricahua city," with able diagrams of "J. Chihuahua, Esquire's palatial residence, which flapped dignifiedly in the breeze" (some horse-blankets stretched from the top of a little scrub-oak); and of "the charming suburban home just completed by our esteemed fellow citizen, Judge Nanay" (a wind-break of brush roofed with blankets, about three feet from the ground); likewise "Dr. Kut-le's elegant new villa," (a few rods of muslin, tented over a peak of Aloe-stocks); besides "the beautiful Miss Na-dis-ough's exquisite cottage, flanked with rare tropical growths," (which the same it was a recess in the steep bank, with some sheeting stretched across the front from bush to bush; the skyward selvage soaring aloft on the point of a pole, the lower anchored down with rocks, and a row of mescal-cactus for the rear wall); and so on for quantity. It was

AN INTERESTING BURG,

and no mistake. Against every bush leaned a gun—maybe half a dozen guns. Every male Indian over three feet high wore one, two or three cartridge belts, filled with the overgrown 45-70 cartridges of Uncle Sam's rifles; a butcher-knife of Sheffield make, in a sheath which swallowed nearly all the handle, as well as all the blade; a curious leather sheath in which is transported the indispensable awl for mending the moccasins; numerous bracelets and necklaces of big, gay beads, and generally a small looking-glass and materials for improving the complexion according to Apache ideas. The bronco bucks were very shabbily dressed, as should be expected after the enormous hardships they have endured, and the incredible distances they have made in the last ten months. Their average dress was a dirty print shirt, whose appendice

TOYED IDLY WITH THE BREEZE,

and they had to do a good deal of toying, for the breeze here for the last week has been one to sweep all the sand in Arizona over into New Mexico. Somewhere under the tailness of the shirt, begins a pair of stout linen drawers, designed for white, but of a present color which is a monument to the vanity of earthly hopes. These draw closely around the ankles, so as to go inside the extremely tight-fitting moccasins. As the Apache perambulates in a cactus country, he cuts his moccasion accordingly. The toe, instead of coming to a flat point like a piece of pie, is strongly pug-nosed. Not only is the point turned up, but it has a little circular shield at the end, perpendicular to the ground, and reaching two inches above it. This disc is as big around as a quarter, and serves as a high protective tariff against the importation of foreign thorns. The moccasin is not of shoe caliber, like those of the Navajoes, Pueblos and other Northern tribes, but has a leg about thirty inches long. The top of this is turned over, so that the business reaches to within three or four inches of the knee-pan. The assassinating cactus, which goes through ordinary shoe leather as 'twere a pleasure jaunt, is stumped by this elastic buckskin. But the linen unmentionables do not suffice for the modest buck. Every one of them has an external and

GENEROUS G-STRING,

whose extremities form small aprons in front and rear. In fact this is a *sine qua non.* Those who chafe at the restraining of the under-pants don't have to wear 'em; but without the G-string no one can hope to be admitted to polite Apache society. A good many robust bronchos—and scouts, too, for that matter— have been parading about the camp and post in bare-kneed pantlessness. The head-dress most in vogue is a big bandanna rolled to a band of two inches, and tied about the head to keep the long, black hair from eye-tickling. In the evening, or when riding in the cold, a blanket or patchwork quilt is added to this costume. Chihuahua and a few others have once-gaudy Mexican serapes. The women wear calico dresses of uncertain denomination, but all modest. The Mother Hubbard seems to prevail only among the old women. The girls don't wear it—so Baldwin and his blushing Coltonites[7] could come out here without any danger of being shocked. The moccasin is similar to the men's. The female children are women on a small scale, so far as dress goes; but the boys, from little to big, run around in the airy costume of a shirt and G-string. Little fellows of seven or eight carry their cartridge-belts and big knives; and one eleven-year-old had, when captured, a gun and a six-shooter. Master Apache gets in his apprenticeship at piracy very early, you see.

All of them are great hands for ornaments, but are not disposed to be piggish about it. All the scouts and most of the broncos had some silver trimmings; but when silver ran short, other metals would do. The silver goes in Navajo style on their hats, belts and necklets. All their metal bracelets were brass, rather cleverly worked, and so were most of their rings. They go on the principle that

A FEAST IS GOOD AS ENOUGH,

and "while they're gittin', they git a-plenty." One coffee-cooler whom I watched gambling, had 15 rings on his left hand, 11 on his right, and a dozen

---

[7]The reference probably is to E.J. (Lucky) Baldwin, a flamboyant southern California real estate entrepreneur of the day. Colton, east of Los Angeles, was enjoying a land boom at the time and would be incorporated the next year, in 1887.

bracelets of beads and brass on each wrist. He was probably afraid that if there were but one or two, they might get lonesome. A good many of the girls supplemented their brass bracelets with tin ones made from old cans.

In looks, these people are not particular facial scarecrows. In fact they are good-looking Indians. Chihuahua has as pleasant a face as one would care to meet—strikingly good natured, and very intelligent. One or two have bad faces—but I can pick you out worse on the streets of Los Angeles, any day. There were five or six rather comely girls in the outfit. Old Nanay, a chunky, fat, superannuated chief, might readily be taken, so far as face goes, for a wealthy Mexican ranchero; and he is facile with a somewhat corrupted Spanish.

Wildest in the rough sports of the broncho boys was one figure which you would single out at a glance. His sandy hair cropping out under a dirty cotton rag; his light skin, pretty liberally exposed and everywhere a mass of miscegenated dirt and freckles, showed that he was no Chiricahua. He was

THEIR LITTLE WHITE CAPTIVE,

Santiago McKinn. This poor child, scaly with dirt, wild as a coyote, made my eyes a bit damp. His is a pathetic case. One day last summer, Geronimo and his band swooped down upon a little ranch on the Mimbres river, above Deming, N.M. They did not attack the house, but skimmed along the range, where the two boys were herding cattle. The elder was killed, as nearly as we can learn, and Santiago, now 11 years old, was carried off. He has been with the Apaches ever since. But this is not what seems so pitiful. He has had to share their long marches, their scanty and uninviting fare, and all the hardships of such a life, no doubt; but he has not been maltreated. The Apaches are kind to their children, and have been kind to him. The sorrow of it is that he has become so

ABSOLUTELY INDIANIZED.

It was almost impossible to get hold of him in camp. The Indian boys liked to be talked with; but let a white man approach, and Santiago would be off instanter. He understands English, and Spanish (his father is Irish, his mother a Mexican), but it was like pulling eye-teeth to get him to speak either. Yesterday Gen. Crook had perfected arrangements to have him taken home. He utterly refused to budge toward the fort with any white man, and Chihuahua had to bring him up. The Apachefied child was brought over to Major Roberts's house. Gen. Crook, Major and Mrs. Roberts and some other ladies were sitting on the porch, and were greatly interested. When told that he was to be taken back to his father and mother, Santiago

BEGAN BOO-HOOING

with great vigor. He said in Apache—for the little rascal has already become rather fluent in that language—that he didn't want to go back—he wanted always to stay with the Indians. All sorts of rosy pictures of the delights of home were drawn, but he would none of them, and acted like a young wild animal in a trap. When they lifted him into the wagon which was to take him to the station, he renewed his wails, and was still at them as he disappeared from our view. By this time he is probably at home. I hope he is finding the welcome that a good home would give to such a return.[8]

OFF TO PRISON.

All this morning the bronco camp was a scene of confusion. The bucks were greasing up their hair, and gathering their cartridges. The squaws were racing over the hills, catching and saddling the mules and horses, and packing cleverly upon them the blankets, muslin "tents," pots and cups, canteens, baskets, and hunks of jerked meat. Chihuahua, with his bright seven-year-old boy clinging behind, was riding up and down all the time to hasten idlers. At last, at 11:30, the camp was packed and afoot. A queer procession it was that wound down Apache Pass and out upon the dusty plain. Here was a gaily painted scout wearing the army blouse, and with his rifle or carbine across his saddle. Beside him, perhaps, was an equally painted bronco, equally well armed, but without the blouse. Next you might have seen a burro so hidden by big bundles that only his slender legs and comical head were visible, while on top and bestride of the whole aggregation would be a squaw, with the peculiar Apache cradle under one arm and across her lap, while the other hand was occupied with whip and bridle. One little pony carried a big buck, a solid squaw and a cradled baby. General Crook and Major Roberts's little sons accompanied the procession with a buckboard. The station was safely reached, the strange passengers were loaded into the emigrant sleepers, and now are trundling eastward, to be stared at by tens of thousands of eyes within the next few days.

LUM.

———

Los Angeles Times, April 13, 1886:

FROM FORT BOWIE.

———

Gen. Miles Arrives and Assumes the Command

[SPECIAL DISPATCH TO THE TIMES.]

FORT BOWIE, A.T., April 12.—Gen. Miles arrived last night, and took command. Gen. Crook issued orders relinquishing his command, and thank-

---

[8]Bourke gives some additional details on McKinn in *On the Border With Crook,* 477.

ing all the officers and men for their zeal, intelligence, energy and courage throughout the discouraging campaign. He leaves this noon for Whipple Barracks, and thence in a few days for Omaha. Gen. Miles says there will be no immediate change in the campaign.

LUM.

CHAPTER IV

Los Angeles Times, April 13, 1886:

A BOWIE BUDGET

Reason Why General Crook was Relieved.

POW-WOWS WITH THE CHIRICAHUAS.

*Chihuahua's Serial Oratory—Crook Bids His Scouts
Good-Bye—Honesty the Best Indian Policy, Etc.*

[STAFF CORRESPONDENCE OF THE TIMES.]

FORT BOWIE, A.T., April 10, 1886.—Gen. Miles is expected here to-morrow, and Gen. Crook intends to leave on the following day to take his new charge—the Department of the Platte. That he is relieved, not only officially, but mentally and physically as well, I know. The long, strong tension, the ceaseless vigilance, the continual grasping after the ungraspable—all have told on him. They would tell on iron. The Indians bore him to death, too. Scouts or captives, they hang around his quarters like flies on a molasses keg. Their most trivial complaints, wishes or hopes, they insist upon confiding to him; and, with his gigantic patience, he listens to it all as kindly as a father could to the endless prattling of a child. But it is tiresome enough, you may be sure. Several weeks ago he wrote to Washington, asking to be relieved. A few days after I got here, the telegram came granting his request. Meeting him in his office, I said I was very sorry to learn that he was to leave. He said, promptly and earnestly: "Well, *I'm* not. I have had to worry along with these fellows for eight years, and have got enough of them. Now, let some of the others try their hands." It is worse than foolish to construe his application for relief into an admission of failure, as some Territorial alleged newspapers have done. It would not surprise me at all if Geronimo and the rest of his band were to be brought in by Gen. Miles shortly—but

WHO PAVED THE WAY?

Who knocked the strength out of the renegades? Who has sent to a Florida prison the most audacious and bloody raiders of them all, and three-fourths of their whole band? Who has chopped off the Apache hydra's chief heads— Ulzanna, Chihuahua and Kut-le; leaving only Geronimo, one good head, and Natchez, a dough-head? Why, old Nanay, fat, aged and lazy as he now is, was more force to the renegades than Natchez. If properly managed, the whole matter can be wound up very soon. But if a stranger comes in and fails to handle these unique savages with all the skill of a man who has dealt with and known them personally for fifteen years, will he be in the least to blame?

General Miles, who has a superb reputation as an Indian fighter, comes here under a vast disadvantage of unfamiliarity with the Apaches, their unparalleled wilderness, and their unparalleled methods of warfare. He will do his level best, there is no doubt of that; and I hope, however his efforts may result—that he will be given the fair play and honest treatment which have been denied General Crook.

A POW-WOW WITH CROOK.

By the way, I have never told you yet about the pow-wow on the day when the prisoners came in with Lieut. Faison. It may interest you, as it did me. We were all watching, that day (April 2), for the arrival of the much-expected captives. At last, just after noon, a grotesque figure came out against the sky over a high ridge on the Bear Springs trail from the south. It came down the rough slope toward us, and soon another popped above the horizon, and another, and another, and so on for half an hour. Lieut. Faison and Dr. Davis rode at the head of the straggling procession, and the rest came as it chanced, bronco bucks, scouts, women and children, sometimes within a rod of each other, and sometimes with a hundred yards of interval. They rode up to headquaters, stopping only for a brief greeting, and then on down to the arroyo to make the camp, of which you have already heard. Later in the afternoon, Chihuahua came up with Alchesay and Keowtennay (he spells it Ka-e-te-na, but I have adopted the phonetic spelling, as Apache has no rules), for a talk with Crook. Alchesay and Keowtennay, you will remember, are two friendly Apache chiefs—and valuable, indeed have been their services in this campaign. The conference, or confab, was held in Gen. Crook's private office. Present, the General, in a rawhide-seated chair; Major Roberts, his Adjutant; Keowtennay, in another cow-chair; Alchesay, on the floor next to my camp-stool; Chihuahua, squatted against a side door, with his knees up and touching, and his feet apart; José María, an old Mexican, who *sabes* only his own language and Apache; and Mr. Montoya, a Mexican, who talks good English. It was a

ROUNDABOUT CONVERSATION.

Gen. Crook would ask a question in English; Montoya would put it into Spanish; María snatch it up and twist it into Apache; Alchesay or Keowtennay would answer, and it went through the reverse process till it came out in plain United States.

In answer to questions about Geronimo's stampede after the surrender, Keowtennay said that the first unpleasantness began the night before the surrender. They had got hold of Tribolet's whisky, and were filling up. Some one flirted with one of Nachita's squaws, Natchita grew jealous, quarreled, and shot the woman through the knee with his six-shooter. On the night of the stampede, Chihuahua, Keowtennay and Alchesay camped close to Lieut.

Maus; while Geronimo, Nachita and their immediate followers settled upon a hill a short distance off. None of them knew of the escape till next morning. The fugitives took so few women, because more would burden their flight. Nachita would undoubtedly get lonely, as he is fond of a good many fat young squaws.

Chihuahua said he thought it very likely that

NACHITA WILL RETURN,

though it was a *quien sabe* case. He didn't believe, however, that Geronimo would ever be seen again. He said the one fat young squaw Natchita took along was very good looking. "I haven't seen Mangus since I left the reservation. Mangus went off it with Geronimo, but they soon quarreled and separated, and didn't meet again till last rainy season [August]. Then they just met and parted, and haven't seen each other since. Mangus had seven in his band, and one of them was killed in Mexico. No, Geronimo wasn't wounded last summer [it was reported that he *was*]. Kut-le was, though, and is still lame."

After Gen. Crook had asked all the questions he wished, and was about to dismiss the meeting, Chihuahua hitched forward several feet toward the General, and said he wanted to talk a little.

CHIHUAHUA HOLDS FORTH.

"When a man thinks well, he shows it by his talk. I have thought well since I saw you. Ever since you were so kind to me in the mountains [at the surrender] my heart has quieted down. My heart is very quiet now. Geronimo has deceived me as much as he did you. I was very glad when I saw my sons and wife. I think well of my family, and want to stay with them. Those who ran off did not think well of their families, nor show love. I am a man that whenever I say a thing, I comply with it. I have surrendered to you. I am not afraid of anything. I have to die sometime. If you punish me very hard, it is all right. I am very much ashamed that the others ran off, and hope you don't think I had anything to do with it. I am much obliged to you for your kindness. I surrendered to you, and have thrown away my arms. I didn't care any more for my gun nor any weapon [this is true. The old fellow didn't have so much as a knife, all the time he was here]. I surrender to you and ask you to have pity on me. You and nearly all your officers have families, and think very much of them, as I do of mine, and I want you to remember your families. I hope you will not punish me very hard, but pity me. I am very grateful to you. I have been very happy since I saw my family again [they had been prisoners here for some months]. I was sleeping very quiet and happy with my family at Camp Apache [at the reservation], but Geronimo came and deceived me, played me a trick and made me leave. Wherever you put me, keep me away from Geronimo and his band. I want nothing to do with them. People will talk bad about me and get me into trouble. I don't want anything to do with him. I was

62

very quiet and happy at Camp Apache, looking after the little crop which I had in the ground, and my horses and wagons, but Geronimo came along and told me so many lies that I had to go [this is strictly true]. It is true we have stolen many cattle and horses, and done many depredations, but Geronimo is to blame for all we did. Now I have surrendered to you, I am quiet and happy. It is very good to see my children, and I want to live happy and quiet all the time. That is all."

The old man has a very kindly face and musical voice, and it was interesting to hear his speech, which took him, with the necessary interpretation, about fifteen minutes.

ANOTHER TALK.

There have been many other little pow-wows between Crook and the Indians here, but none of particular interest until last Tuesday, the day before the prisoners were sent off. That afternoon Noche,[1] Dutchy (who killed the slayer of Capt. Crawford),[2] Stovepipe, Charley,[3] and other scouts were squatted upon the porch at headquarters. The interpreters sat upon chairs, and the General occupied another, tilted back against a window. He was tired and preoccupied. The Indians had been chattering at him for a couple of hours with their childish requests. Most of them wanted letters written to their friends on the reservation, telling when they'd get home. Others had directions to send about the care of a horse or a hoe. Crook listened patiently and arranged their little perplexities. Then he gave them,

A LITTLE VALEDICTORY.

He has hitherto kept the knowledge of his removal secret from the Indians, lest they should get uneasy. He said to the scouts:

[1]Noche (1856–1914) was sergeant-major and a top scout with Crawford during his Sierra Madre operations. He was exiled with the rest of the Chiricahuas but eventually returned to the Mescalero Reservation in New Mexico, where he died.—Griswold.

[2]Dutchy was one of the most famous of Apache scouts, largely because of a persistent, lengthy effort on the part of Arizona civil authorities to get custody of him to answer for alleged offenses against the white community before he turned scout. As Lummis reports, Dutchy is said to have killed the unidentified slayer of Crawford. (This was not Mauricio Corredor, who had wounded Tom Horn and who also was killed that day by a scout; Corredor was considered by some to have been the individual who had killed Victorio five years earlier). On a later occasion Dutchy saved the life of Lieutenant Britton Davis. Dutchy was killed in Alabama in 1892.—Griswold; Thrapp, *Victorio*.

[3]Stovepipe first was enlisted as a private in Company A of the Apache scouts October 1, 1879, by Second Lieutenant James A. Maney, for service in New Mexico. Charley also was enlisted as a private at that time and was re-enlisted by Crawford December 25, 1882, for scouting duty along the Mexican border, apparently going into the Sierra Madre with Crook. Both apparently were enlisted in November, 1885, for Crawford's final Sierra Madre expedition.—Indian Scouts Muster Rolls, Records of the United States Regular Army Mobile Units, 1821–1942, Indian Scout Companies 1872–93, Record Group 391, National Archives.

"I am going to leave you. Another officer is coming in my place [sensation in the audience]. I want to thank you for the good work you have done. You have been very faithful. I have made many enemies among my own people by being honest and square with you. After I am gone, probably some will tell you lies about me. But you must judge me by my acts. Talk is cheap. I hope you will remember the good advice and teachings I have given you. Do everything to stop this tizwin-drinking. You get it in your stomach, and there is no sense left. Then you go and gamble away all your money, and the next day you have nothing to show for it but a swelled head [laughter]. Go to raising stock, as well as the farming you have already learned. You will do better to raise sheep than cattle. To get anything out of a steer you have to kill him; but you can sell the wool off of your sheep, and still have the sheep [approbative chorus of "Hu! Hu!"]. Then the sheep, though they have to be watched, will not wander off to great distances, as cattle will ["Hu! Hu!"]. Besides, you know how thieves come and run off a few of your cattle, and it is hard to get them again; but the sheep, when a few go, all go, and travel very slowly, so you can easily catch them and the thief [very emphatic "Hu! Hu!"]."

THEN CHIHUAHUA,

who had been sitting at one side, nudged and crawled over by his bright little boy, put in his oar as follows:

"I am glad that you will talk to us. As soon as I saw you, down there in the mountains, it seemed to me that I quieted down very suddenly. My heart got very quiet. After I saw you, I slept well that night, and drank nothing but water. It seemed as if I had suddenly got well from some disease. Since I surrendered to you I have been very quiet and contented. You can do a great deal of good to me. I am very glad, because all you tell me is right. I am very much satisfied to be with my family. I believe every word you tell me. Wherever you send me, it is all right. My children cry a great deal to think of leaving two horses I have, and I would like to take them with us. I would like to know when you are going to send us away, so I can collect some money the scouts owe me."

Gen. Crook: "Well, you'd better collect it right away."

Chihuahua: "We would like to have an officer go with us."

Gen. Crook: "I will send a good one with you, and one that can talk Spanish."

Chihuahua: "I am satisfied and content, wherever you send me. I think all my people will behave all right, and I hope sometime to come back and see my people. They gave me a good wagon up at the reservation, and I am afraid that if I am gone a very long time, I cannot get that wagon again when I come back."

Gen. Crook: "Oh, I'll tell your relatives to take care of that."

64

ABORIGINAL HONESTY.

Poor old Chihuahua said "thank you," and trotted off, evidently feeling blue. His talk indicates the *naive* childishness of his race in some points, rugged and self-reliant as they are in others. The guarantee of his utter honesty lies in his acts. He could have absolutely escaped with all his people a hundred times while they were in camp here. They could have escaped en route to the train. But they didn't make the slightest offer of it, though they had a good many forebodings as to what will be done with them. So far as this honesty was concerned, they could safely have been sent to Florida without a single guard. The soldiers were necessary to protect these poor savages from the "civilized" whites along the way. Why, Gen. Crook wouldn't let me telegraph that they were going, until the night before they went, for this very reason. There are plenty of alleged white men who would jump at the chance to signalize their bravery by shooting a captive squaw through a car window, if they had received sufficient notice to brace themselves with brag and whisky. Here, too, you can see the advantage of honest dealing with an Indian. These people

TRUSTED CROOK,

and faithfully, in their turn, kept their pledges to him. It was very fortunate, for all the troops at Fort Bowie and Bowie Station wouldn't have been enough to disarm and bind those Apache prisoners. In fact the history of the Chiricahua "captives" from the surrender till they were put on board the cars with an armed escort, has been that of people on parole, rather than of prisoners. The Territorial papers howled and damned Crook for a fool; but the logic of events sets him clear and clean above their malice. Instead of trying what he knew was impossible: killing a few women and children and losing all the warriors to be [eternally] hunted, he gave his word and took theirs, and "got there." He did have trusty men camped with the prisoners to talk down any dissatisfaction, but that is all. Now let us see how soon anyone else will get hold of 15 warriors, dead or alive—not to count at all the 62 women and children! The remark made by the scouts when Crook told them of the official change is significant. They said

"WE DON'T KNOW THIS MAN

that is coming instead of you. What is he? Is he a good man, or will he lie to us the same way other people have lied to us? We would like you to write us a letter telling him that we have done right, and to do right with us."

Gen. Crook told them that his successor would be a good and honest man, and promised to recommend them to him. The Apaches have been lied to and swindled so that they don't know whom to trust, and are very suspicious of any man until they have tested him and found him honest metal, as they have Crook.

I mentioned, the other day, the stubbly chin swath in Gen. Crook's fine full beard; and now I have learned the cause. He was loading some paper shells, some weeks ago—for he is as mighty in the hunt as on the warpath—when one of them exploded in his hands: cut an ugly hole up through his beard, and burned his eyebrows nearly off. Luckily, the shot in the cartridge failed to hit him; but his face was badly burned. It was a close call.

LUM.

Los Angeles Times, April 15, 1886:

GERONIMO

At His Old Tricks—He is Joined by the Apache Scouts.

EL PASO, April 14.—An American gentleman, who has just arrived in El Paso from Western Sonora, reports that Geronimo and his band of hostile Apaches are doing considerable mischief near Bavispe. There seems to be very little doubt that nearly all the Apache scouts that were in the United States service during the recent Apache campaign under Crook have joined Geronimo. These same scouts, while still in the Federal service, last winter, committed many depredations in that portion of Sonora, stealing stock on all occasions, and when ordered to return the same by their commanding officers, they flatly refused to do so. It is thought that in the case of many of these scouts fear of future punishment has caused them to join Geronimo.

Los Angeles Times, April 16, 1886:

AT THE FRONT.

Crook and Miles Among the Indians on the Border.

THE VETERANS EXCHANGING COMMANDS.

Scenes at the Greeting and Parting—Warm Tribute to the Gray and Gallant Campaigner of Thirty Years.

[STAFF CORRESPONDENCE OF THE TIMES.]

FORT BOWIE, A.T., April 12, 1886.—Yesterday afternoon the boom of the six-pounder, eleven times promulgated upon the air, notified us that the ambulance containing the new commander of the Department of Arizona had rounded the bend in the road. In a few minutes more the six mules swung around the corner of the store, and trotted smartly up across the sloping parade, and drew up in front of Col. Beaumont's house. Gen. Miles crawled out of the inadequate door of the ambulance, and shook the kinks out of his legs. Gen. Crook walked up from the office, and the two veterans shook hands undemonstratively. After dinner with Col. Beaumont, Gen. Miles came down to the office, and passed most of the afternoon in a close conference with Gen. Crook, who explained the situation fully, in response to his successor's questions. In the early evening Gen. Crook issued the following general orders:

> "HEADQUARTERS, DEP'T OF ARIZONA,
> In the Field,
> FORT BOWIE, A.T., April 11, 1886.

General Field Orders, No. 3.

In obedience to the orders of the President, I hereby relinquish command of the Department of Arizona. In severing my official relations with this Department, I cannot but express my appreciation of the zeal, intelligence, energy and courage which have marked the conduct of the troops of my command under the trying and discouraging circumstances attending the operations of the last ten months. For their steadfast loyalty and hearty co-operation, I tender them—officers and men—my sincere thanks.

> George Crook,
> Brigadier-General, Commanding.

Gen. Miles issued a brief General Orders No. 4 in which he assumed command of the Department of Arizona, including the troops in Arizona, New Mexico, at Fort Bliss, Texas, and Fort Lewis, Colorado.

This morning there was another pow-wow in the office, Gen. Miles gathering up all the points as to the situation with which he is now to wrestle. While he and Gen. Crook were talking, the Apache scouts came over to take their leave. The most prominent and valued ones—Noche, Charley, Dutchy, Stovepipe and others—trotted into the office with old Concepción and Lieut. Maus. They did a good deal of talking, and also received a short speech from Gen. Miles. He reminded them how much better off they are than the renegades now lying among the mountains of Mexico, and than the prisoners now in Florida. He informed them that every last one of the renegades will be hunted down and taken alive or dead, no matter how long it takes nor how much money it costs; and bade the scouts behave themselves when they got home.

It was a sight to see the scouts when they came to say farewell to Crook. The common "coffee-coolers" (as they call inferior, lazy scouts) merely shook his hand very effusively, and said good-bye several times over. But the men whom he had trusted, and who had proved their wonderful efficiency in this campaign, they were not content with that—they

HAD TO HUG HIM.

Noche and Charlie, particularly, threw their arms about him, and kept them there some time, patting him affectionately on the back with the right hand. He smiled at them indulgently, and told them to be good Indians.

This noon the scouts left us, steering their mules, horses and burros toward Camp Apache, where they will be disbanded. Mr. [Sidney R.] DeLong,[4] the storekeeper, got back late last night from a long raid through Mexico after a lot of the bright-hued serapes for the scouts. The store was a busy place this morning, and by midday the majority of the scouts were clad in flaming colors that would have knocked Solomon silly.[5] All of them had plenty of cash, having just received four months' pay, and they made it fly. One, who had annexed to his store the proceeds of a little judicious koon-kan, went out this morning and bought three extra horses for $166. He had $250 more to go on, too. There was a funny little incident just before they started off. A white fellow had sold a horse to one of the scouts, "Shorty,"[6] who found the animal a scrub, and yearned to get his fingers again on his $40. He found the seller at the store, and demanded his coin. White man requested him to go to sheol. Scout said, "Come out, look horse. Dam no good." White man came out, and said he "didn't see no flies on that 'ere hoss." He started back to the store, but Shorty collared him, and beckoning at his pocket, said, "Come, money; come, money." White man tried to get away, but Shorty held on. At last the fellow caved, and paid over the $40 with ill grace, remarking, "I'll meet you somewhere else, one o' these times, an' then I'll fix you!" The surrounding scouts roared in derision, well knowing what sort of show the white man would stand in trying to "do up" one of them. Shorty smiled sarcastically, and said, "all right—when you want fight me?

"YOU CATCH HIM, YOU HAVE HIM,"

which was as pithy a *defi* as could be desired. It just about fits this whole matter of fighting the Apaches, too—"You *catch* him, and you can have him."

[4]DeLong (1828–1914) was a prominent pioneer, merchant, and newspaper figure in Arizona. Born in New York, he studied civil engineering, went to California in 1849, was an officer in the First California Infantry during the Civil War, and initially visited Tucson in 1861. Mustered out in 1866, he returned there. In 1871 he took part in the Camp Grant Massacre, later confessing that it was "the only thing he regretted in his life." He was mayor of Tucson in 1872 and after the turn of the century was a longtime secretary of the Arizona Pioneers' Historical Society.—James H. McClintock, *Arizona: Prehistoric, Aboriginal, Pioneer, Modern,* 3 vols. (Chicago, S.J. Clarke Publishing Co., 1916), III, 373.

[5]He means Joseph, of course.

[6]No mention was found of this scout, but the record is incomplete.

With the scouts, also departed Lieut. Marion P. Maus, who was left in command of this battalion of scouts by the murder of Capt. Crawford. He is a very intelligent, square and manly young officer, and has furnished me with what you [may] find suggesting reading, in due time. Second Lieut. W.E. Shipp,[7] of the same command, went off too, but not without similarly crowding a good many pages of my note book. He is a bright and pleasant West Pointer, with the figure of an Apollo.

CROOK'S DEPARTURE.

At 1 p.m. the ambulance and its sextette of long-ears was drawn up in front of Major Roberts's door. Several unassuming bundles in canvas were strapped behind, and then an old linen duster appeared in the door—and Gen. Crook was bidding farewell to the group on the piazza. If ever cordial good wishes followed a man from the scene of his labors, from those who were in a position to know him, they followed that one. When the doings of this decade have been refined from prejudice into history; when the mongrel pack which has barked at the heels of this patient commander, has rotted a hundred years forgotten—then, if not before, Crook will get his due. In all the line of Indian-fighters, from Daniel Boone to date, one figure will easily rank all others—a wise, large-hearted, large-minded, strong-handed, broad-gauge man—George Crook. I am glad that, after a third of a century of almost steady campaigning, he has at last secured a needed and nobly-earned rest, and goes to take life quietly in the department of his choice (as ranking Brigadier of the army he *had* his choice, and selected Omaha). He is now on his way to Whipple Barracks, A.T. (to reach which he has to go around by Albuquerque and out on the A. and P. [Atlantic and Pacific Railroad]), where he will remain a few days to pack up; and then he will proceed to his new post. May all success and prosperity attend him.

LITTLE LAURA.

I have already mentioned his fatherly kindliness to the Indians, in all their tediousness and childishness. It touched me a good deal. Equally noticeable is his fondness for children. In the busiest and most discouraging hours of the campaign he would unbend his eyebrows for a moment's romp with little Laura, Major Roberts's lovely three-year-old girl. She thinks the world of the tall old warrior who will drop the Apache renegades for a space, to toss her in his arms. She calls him "Crookie," with cheerful unconventionality, and has no compunction about assailing him at any point. Her bugaboo is the goat,

---

[7]William Ewen Shipp (c. 1861–1898) was a North Carolinan who became second lieutenant in the 10th Cavalry upon graduation from West Point. He was killed July 1, 1898, at the battle of San Juan, Santiago, Cuba.—Cullum; Heitman.

which she holds in holy horror. When that exploding cartridge mowed its way through the General's beard and eyebrows, and burned his nose and cheeks, Laura was present. She set up a wail of distress, and ran to her mother, saying, "I don't like Crookie any more—he looks like a billy-goat!" Ever since that catastrophe she has refused to sit next him at table. She rolled me over, the other day, when the three of us were sitting upon the porch, by exclaiming, when the General wouldn't tell her something: "Crookie, you're an old bump on a log!" Verily, childhood, like death, is no respecter of persons.

By the way, Major Roberts and family will spend some time in and around Los Angeles, shortly, taking advantage of his leave of absence. The Major has served on Gen. Crook's staff at intervals ever since the beginning of the Civil War, and is very highly valued by him, both as an officer and as a friend. General Orders, No. 4, relieving Major Roberts from further duty here—and which he was too modest to let me see—is one of the warmest recognitions of faithful service I ever read on the sly.

This afternoon I had a short

TALK WITH GENERAL MILES.

He is a tall, straight, fine looking man, of 210 pounds weight, and apparently in the early fifties as to age. He has a well-modeled head, high brow, strong eye, clean-cut aquiline nose, and firm mouth. It is an imposing and soldierly figure, all around. He said: "There is no change. Capt. Dorst is still down in Mexico, pushing after the hostiles, and it will be some little time—say a fortnight—before any additonal operations can be organized. We shall simply keep up the pursuit already inaugurated, always following the hostiles and giving them no respite. Thus we shall serve the double purpose of the campaign—to protect the settlements and get hold of the hostiles. The pursuit will be kept up until we get them, dead or alive." He also asked how Los Angeles was prospering, and highly complimented its beauty. Having, in the Washington papers, an intimation that the Indian scout policy is to be discontinued, and white scouts employed, I asked Gen. Miles about it. He said he was not yet prepared to declare his plans in this respect. I have no doubt that if it is left to him, he will take the Apache scouts. It has been his policy hitherto to employ Indian scouts; and he has already been here long enough to understand, no doubt, what an absolute farce it would be to try to catch the renegade Apaches with white people. But there are indications that the President has again exposed himself, and is inclined to run the army in detail. If so, it is a remarkable case of ill taste and ill sense. The idea of a man running this campaign from Washington—a man who never saw the frontier, and who knows less about Indians and the details of such a war as this than the dullest officer in the field! But it's just like Cleve.[8]

LUM.

Los Angeles Times, April 17, 1886:

CROOK'S APACHE SCOUTS.

No Truth in the Report that They Have Joined Geronimo.

[SPECIAL DISPATCH TO THE TIMES.]

FORT BOWIE, A.T., April 16.—The Press reports from El Paso stating that many of Crook's Apache scouts in the recent campaign have joined Geronimo, and that while in the Federal service they deserted in Sonora, are Mexican inventions to excuse the murder of Captain Crawford, and are entirely false. All of the scouts who are discharged are now on the reservation, and all others are with their respective command. Far from being reinforced, Geronimo's people are so discouraged by the tireless pursuit that even his little band is disintegrating.

LUM.

[8]President Grover Cleveland. Sheridan, who was responsible for the substitution of Miles for Crook at this point, was a disbeliever in the value and loyalty of the Apache scouts, and Miles had come to Arizona with the understanding that he was to play down the use of the Indians and in their place make prominent—and hopefully effective—employment of the soldiers of his command.

CHAPTER V

Los Angeles Times, April 17, 1886:

SCOUTS AND LIARS.

The Quality of the Animal in Arizona.

HOW "NEWS" IS MANUFACTURED THERE.

Moccasins and Boots—A Savage with a Callous Sole—
Some Views of General Crook on Causes of the Trouble.

[STAFF CORRESPONDENCE OF THE TIMES.]

FORT BOWIE, A.T., April 14, 1886.—If Providence had particularly laid itself out to back up, by sample, what I have said about Arizona economizers of truth and the way Apache war news has been propagated, it couldn't have done much better than in some of the developments of the last few days. The other day I noticed in a 'Frisco paper a long special from Tombstone, relating that "the celebrated Frank Leslie,[1] better known as Buckskin Frank, has just arrived from Sumplaceruther. He has been for many years chief of scouts, is in Gen. Crook's confidence, and gives the first authentic account of the escape of Geronimo." Then followed a lot of porcine-ablution, inaccurate and foolish but harmless. The real news had been published in the same paper several days before, and the above was of course merely a dodge of the alleged celebrated scout to advertise himself. There is a lot of such stuff sent out over the country for no other end than the glorification of

SOME ARIZONA SALOON CELEBRITY.

This man Leslie is a peculiar case—one of the types of a class not infrequently met on the frontier. A man apparently well educated, gentlemanly and liked by all who know him; with as much "sand" as the country he ranges— but a novelist who can make a little truth go as far as any one in the Territory. As much fact as you could pick up on a pin point would last him a year. But there is one thing about it, his prevarications are all harmless. He never lies to hurt anybody—and least of all to hurt Frank Leslie. It is the prime ambition of his existence to figure as a scout—and a scout he will be, if wild-cat dispatches from Tombstone can make him one. He was for a few weeks connected with

---

[1]Leslie (c. 1842–c. 1922) was one of the authentic characters of frontier Arizona. Probably born at Galveston, Texas, he reached Tombstone about 1877, was involved in assorted shootings, sentenced for a time to Yuma Territorial Prison and probably died in the San Francisco Bay area.—Colin Rickards, *Buckskin Frank Leslie: Gunman of Tombstone* (El Paso, Texas Western Press, 1964); Thrapp, *Dictionary of Frontier Characters.*

Capt. Crawford's command, hunting Geronimo, but was directly discharged because of his inability to tell a trail from a box of flea powder. Therein lies his claim to distinction as a celebrated scout. But though no scout, he is no dude. He has killed his two men, under circumstances of Arizona propriety, is a fine shot, and can ride farther and harder in a day than any other white man you can rake up with a fine-toothed comb. As to his "enjoying Gen. Crook's confidence," I guess it isn't necessary to say anything—but you ought to have heard the quiet old General laugh when I showed him that dispatch. Well, so much for that sort of news-fodder.

THIEVISH TICKLERS.

The other instance is of a more aggravated sort, though it has its equally ludicrous side. These telegraph operators out here would steal the coppers off a dead man's eyes and then kick because they weren't half-dollars. They have been divulging secret department news in butchered shape, right along, and now I find they are stealing my dispatches right along. You remember I sent a special on the 7th, telling how the Apache prisoners were shipped to Florida, how they "went cheerfully, though understanding what is to be done with them"; that "two weeks ago, ninety-two hostiles were on the war-path"; and so on. Yesterday I picked up a *Globe-Democrat* and found that identical dispatch dated Bowie Station, April 8. Some larcenic operator stole it and sent it off, but the way he mangled its poor remains was a caution to snakes. He called Nanay "Nona," Ulzanna "Alseanus"; Natchez's children "the Natiche children"; and a lot more of the same. But he is fairly entitled to that whole bakery for saying "all *wept fearfully, not* understanding what is to be done to them"; and "two weeks ago, ninety-two *hostages* were on the war-path!" I hope there is a hereafter!

SOMETHING TO BOOT.

The comfortable condition of my hoofs reminds me of some other hoofs which have my sincerest sympathy. Coming out here, I brought along a pair of boots with blue tops, expecting to do considerable riding. The plaguey things were as sawful on the ankle as if some one had been polishing me with a rasp. So I took them down to the Apache camp, one day, while the whole outfit was there. Struck a nest of scouts and bronchos playing koon-kan, and at once made commercial overtures to one young Chiricahua who sported a fine pair of pug-nosed moccasins. What with two words of Apache, 'steen of Mexican, and abundant United States, I impressed upon him the gorgeous advantages of those boots—particularly the blue-toppedness. He got mashed on them himself, and pulling off a moccasin inserted his toes into the boot. I gave up on the trade at once, for when his toes touched bottom his heel was still half-way up the leg. But he was no tenderfoot. He stamped and pulled and tugged and wrestled, remarked that the boot was "dam not long," grew a fine meerschaum color in the face, sweated and fumed and reared. But he kept at it. I hope I may never see the back of my neck if he didn't toy with that boot for half an hour—and when at last he got it on, his big toe bulged out over the end of the sole. But he was happy. Then he went through the same circus with the other foot, handed me his moccasins, and stomped off around the camp as happy as

A DOG WITH TWO TAILS.

Those boots would have attended the obsequies of any other man save an Apache scout, before this, but a little thing like this won't *wool* him a bit. You see all these scouts are traveling immense distances daily over the roughest country in the world, with nothing between their soles and the rocks but a thin moccasin; and their feet become like the nether milestone or a politician's conscience. I was awfully tickled at the hospital the other day. When the Mexicans attacked Capt. Crawford's force in Sonora, last January, they first fired into the camp, shooting one of the scouts through both hips, and doing him the utmost damage that could be done a man. The poor devil has six holes in him, all made by one bullet. When I went into the hospital to see him, he was feeling pretty blue, apparently and was singing a doleful ululation to get ahead of the doctors. He called on the Earth-mother, the sun, the winds, and various spare entities, to plaster him up and give him a new lease of life. He wouldn't pay any attention to visitors until he had finished his opera. Then he boned me for a cigarette-paper and some tobacco, which he connubiated with Mexican dexterity. Then he sat up in bed, kicked his poor wasted legs from under the cover, turned up one callous foot, and drew one of those torpedo parlor matches

ACROSS HIS BARE SOLE!

A strip of sand-paper couldn't have been more effective, and the match went off like a toy pistol. We all had to hold our sides, for fear of laughing a rib out. No use in expecting that sort of feet to rebel at a little inhumanity on the part of a boot. That fellow, by the way, is fast getting well, and was taken the other day to Fort Apache in a rough wagon.

Speaking of the attack on Capt. Crawford's force, and the cowardly assassination of the brave officer, I see President Diaz has turned himself loose—very loose—on that proposition.[2] It wouldn't be a bad scheme for you to frame that message of his. I have now the full facts as to the murder, and you shall hear them in a few days. Let Diaz keep on writing messages till the day after judgment, and he can't cover the damning facts.

AS TO TRIBOLET.

You will remember that in a previous letter, detailing the escape of

[2]See foregoing, pages 51–52.

Geronimo, the cause of the stampede was given—the selling of whisky to the renegades by a fellow named Tribolet, living in Sonora, just below the line. The fact stands unmoved, but I see somebody has been sending off dispatches to anti-Crook papers setting forth that Tribolet is a very nice man, one of the oldest residenters in the vicinity of Tombstone, wealthy, respected, etc. So I feel constrained to devote a bit more space to showing what this man's lust for gold has done.

Gen. Crook said to me the other day, speaking of the mercenary motives of many people in the Territory: "That man Tribolet is the cause of this whole trouble now. If it had not been for his whisky, those renegades would never have decamped, the whole thing would now be settled, and we would be reaping the results of nearly a year's arduous work. The two bronchos who came in with Lieut. Maus yesterday [the 3rd] tell me they were sleeping together on the night of the stampede. All were pretty full. They had heard nothing of any intention to run, but woke up and saw the others leaving, and went with them, because, being rather tipsy, they thought there must be something wrong. As soon as they found out that there was no trouble, and that it was only a drunken stampede, they came back voluntarily and re-surrender to Maus. If we could get at the others right now, we could disintegrate them, break up Geronimo's following, and then he'd have to come in himself, as he acknowledged at the conference. Oh, no, there's no way of dealing with Tribolet. He has been tried before, but bought his way out. If we had shot him down

LIKE A COYOTE,

as he deserved to be, it would have raised a terrible row. Why, that man has a beef contract for our Army! We are obliged to advertise and let these contracts to the lowest bidder—and he got one. It don't make any difference how big a scoundrel a man may be—that doesn't disqualify him. Punish him by law? We have no laws here! This is a country where the majority rules; and no matter what is on the statute books, no law can be enforced against the sentiment of the community. Now this man Tribolet may be, through that action of his, the cause of the death of a thousand people. There is no knowing where this thing will end, now. And such fellows as he can undo the whole work of a great government, without recourse."

Lieut. Wm. E. Shipp, one of Capt. Crawford's efficient officers, says: "I have known Tribolet a year. When the 'rustlers' were bad, he was notorious as a buyer of stolen cattle. Three or four years ago he was tried for stealing barley from the government, at Fort Huachuca, but he had too much money to be convicted. Last February Lieut. Faison and I went to expostulate with him for selling whisky to our Indian scouts. He said he couldn't be touched for it—had moved below the line on purpose to get away from the law. He did promise us that he would sell them no more, but he kept it right up, and smuggled to the Indians the whisky which led to the stampede."

Lieut. Faison, also of Crawford's command, corroborates this.

Customs Officer Green[3] says: "Tribolet has told me he didn't want the hostiles captured. 'Why,' said he, 'it's money in my pocket to have those fellows out.' He bragged to me how much whisky he had sold them, and how he had given Geronimo a bottle of champagne."

Another custom-officer saw Tribolet sell whisky to the renegades the day before the stampede; and Capt. John G. Bourke saw them intoxicated, Tribolet's being the only place where they could obtain liquor.

U.S. District Judge Barnes said to me: "Tribolet has been a regular receiver of stolen cattle, and is one of the worst scoundrels that ever went unhung."

I could add a lot more such testimonials, but *'sta bastante*. Tribolet has made a lot of money, but it would all melt, one of these days, if he could take it with him.

LUM.

[3]This might have been George M. Green, who as early as 1881 was along the border, complaining that white outlaws were causing unrest in the area.—Henry P. Walker, "Retire Peaceably to Your Homes: Arizona Faces Martial Law, 1882," *Journal of Arizona History,* Vol. X, No. 1 (Spring, 1969), 12.

Chihuahua, noted Chiricahua chief who held several conversations with Crook while Lummis was present. Photograph courtesy the Arizona Historical Society.

John Finkle Stone, Apache Pass miner killed by Apaches in 1869, buried at Fort Bowie. Photography courtesy Fort Bowie National Historic Site.

Santiago McKinn, Apache captive who was freed by Crook, shown in a Camillus Fly
photograph before surrender of the hostiles. McKinn, who had become attached to his
captors, was freed against his will and sent to relatives near Deming, New Mexico.
Photograph courtesy the Arizona Historical Society.

Dutchy, one of the most famous of the Apache scouts, made so by prolonged Arizona attempts to gain civil custody of him to answer for alleged crimes while he was a hostile. Photograph courtesy the National Archives.

Sidney R. DeLong, sutler or storekeeper, at Fort Bowie when Lummis was there. His place was a general meeting ground for Indians and whites. DeLong had a role in the Camp Grant Massacre of 1871 and was a prominent Arizona pioneer. Photograph from the Sacks Collection, Arizona Historical Foundation.

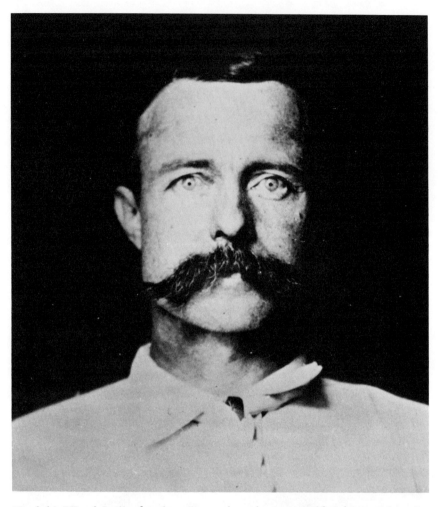

"Buckskin" Frank Leslie, for whom "as much truth as you could pick up on the point of a pin would last a year." A fine fellow, but no Indian scout. Photograph courtesy the Arizona Historical Society.

This is the Fort Bowie cemetery of 1886 as seen and described by Lummis. Photograph courtesy the Arizona Historical Society.

This was the sole marker remaining in the Fort Bowie Cemetery of 1971, its legend undecipherable, the place overgrown by brush and weeds. The cemetery has since been partially restored to resemble that seen by Lummis toward the close of the Apache wars. Photograph by the author.

CHAPTER VI

Los Angeles Times, April 18, 1886:

HEADBOARDS.

An Hour in the Lonely Graveyard at Fort Bowie.

A GHASTLY LIST OF APACHE VICTIMS.

*"Killed by Apaches," "Tortured to Death by Indians"—
a Literary Officer—How Miles Takes Hold.*

[STAFF CORRESPONDENCE OF THE TIMES.]

FORT BOWIE, A.T., April 15, 1886.—Graveyards, as a rule, haven't much of a pull on me. Indeed, they strike me as unsociable and slow. But I strolled through one this morning which is interesting—the burial ground below the Fort. This Post has been here ever since 1862, when it was founded by the California volunteers to keep the Apaches from the circumadjacent springs, in the old overland mail days; and the little graveyard has grown apace since those troublous times. Three-quarters of a mile west of the Fort, on a pretty little bench above the arroyo, it stands within a high picket fence; its white head-boards shining in the sun, while sparrows chatter and doves coo over the narrow mounds. Between its ridges the ground is gay with golden poppies and snowy margueritas; and here and there upon some unforgotten grave, a buckhorn cactus spreads its prickly antlers, or a turk's-head nestles close against the bare gravel. There are but two memorial stones in the whole inclosure; all the rest is lumber—pine boards planed and painted white, while cramped black letters, in straggling lines, tell their terse story. How many a romance is in that lonely half-acre! All burying grounds cover that which was once hope and joy and love; but this little bench along the barren mountain-walls of Apache Pass, is eloquent with the story of the Arizona frontier. It is full of Apache workmanship. The dumb upheavals of its brown breast tell of the old stage creaking through the desolate cañon; the sudden little puff of smoke from behind yon innocent tuft of bear-grass, matched in a sickly curl from that rock and another from the aloe-bunch beyond; the sturdy driver tumbling from his perch; the tangled horses floundering in terror; the ashen traveler dragged from his concealment; and last of all a horrid bonfire, whose odorous smoke goes up with the tortured shrieks of a writhing form.

There are thirty-three graves in this inclosure whose head-boards record simply that—— ——died on the——th of——, 18—. Of these, seven were children. The presumption is that they passed away in the course of nature. There is another board whose weathered face no longer tells of what lies beneath. One of the first graves we reach is that, so the inscription says, of

O. O. SPENCE
Born in Pennsylvania.
Aged 28 years.
Killed by Indians April 7, 1876.

Close beside it is a broad, wooden cross, across whose arms runs, in ornate letters:

NICHOLOS M. ROGERS.
Born in St. Joseph, Mo.
Killed by Indians April 7, 1876.[1]

Next it in the line stands one to the memory of:

JOHN McWILLIAMS
Killed by Apaches Feb. 26, 1872.
Aged 26.

Beyond this is another but wider mound, whose board relates that here lie

A. F. BICE, F. PETTY, F. DONAVAN.
Killed by Indians in Apache Pass, Jan.
24, 1872.

These men were coming up, two on horseback and one on a buckboard, close to this very spot, when the lurking Apaches shot them down.
Next is a board bearing the single inscription:

In Memory of
Col. STONE.
Supposed to be.

What a page of pathos there is in that last line! Col. Stone had a mine on the mountain just back of the Fort. One day he disappeared; and searchers found

---

[1]Spence and Rogers were killed by Pionsenay, a particularly unruly Chiricahua, and their deaths signaled a course of events that profoundly changed history in the region and opened a decade of Apache disorder, wars, raids and deaths. They were shot in a dispute with a small band of Chiricahuas. The incident opened the way for John Clum's attempted transfer of the central Chiricahuas to the San Carlos Reserve, a key move of the so-called "concentration program," which was intended to bring most Apaches to the central reservation. This undertaking, fine on paper but unworkable in practice, proved a disaster for Indian-white affairs in Arizona.

only a sickening mass of meat, hacked beyond recognition, but "supposed to be" the missing miner.[2]

There are other graves of similar marking—"Lieut. Julian Agueira, supposed to be"; "James McIntyre, supposed to be"; "John Kilbey, supposed to be."

I heard the story of another grave, whose board says:

In Memoriam of
GEO. KNOWLES,
Prvt. Co. H, 32 U.S. Inf., captured and tortured
to death by Apache Indians,
May 26, 1868.

Knowles and another private named King were acting as guards to the stage down the pass. Just before they emerged to the plain, the stage was "jumped" by Apaches, the driver being instantly killed. The conductor stood off the foe for a little while, but was soon wounded and overcome. He was known here as "Tennessee." For some reason or other, the Apaches did not kill him, but carried him off a captive. He was afterward killed in Mexico in a fight between the Apaches and Mexicans. The two soldiers with the stage do not appear to have made any resistance. No empty shells from their guns could be found. They probably threw up their hands and surrendered. Poor wretches! Their savage captors bound them and roasted them alive, with every fiendish ingenuity of horror.

In the northesat corner of the yard are two tiny graves, side by side. Their headboards are like the rest in make, but unique in legend. One inscription says:

In memory of
LITTLE ROBE,
Son of
GERONIMO,
Apache Chief.
Died Sept. 10, 1885;
Age 2 years.

[2]Lummis here leapt to the wrong conclusion. The phrase, "supposed to be," did not refer to remains unrecognizable, but to the fact that since the headboards were placed long after some of the graves had been filled, it was not certain they were in the correct position. John Finkle Stone (1836–69) was a prominent southwestern pioneer. Born in New York State he had roved the west from Utah to Arizona and eventually established the Apache Pass Mining Company, a profitable gold-extracting enterprise. The stage on which he was journeying to Tucson was ambushed October 5, 1869, by Indians near Dragoon Springs, and Stone and five others were killed, he being taken back to Apache Pass for burial. Stone Avenue, a principal Tucson thoroughfare, was named for him.—Thrapp, *Dictionary of Frontier Characters.*

The other head-board reads:

In memory of
MARCIA,
An Apache child.
Died July 3, 1885;
Age 3 years.

These little savages, captured with others of the renegades last summer, died here in the guard-house. It was a soldierly and a manly heart which took care that these poor little waifs were decently buried, and that their last resting-place was marked. It was an act of humaneness of which I fear but few Arizona civilians would be capable.

THE REST OF THE ROSTER.

Besides the victims of Apache ball and arrow above mentioned, the following names will be found on the head-boards of this quiet city of the dead—each with the ominous line below: "Killed by Apaches." I append the date when killed, and the place, when mentioned:

Lieut. John C. Carroll, Thirty-second Infantry, November 5, 1867.[3]
Samuel Hickman, private, Troop F, Fourth Cavalry, October 10, 1885.
John M. Coss.
J.F. Keith, June 25, 1862.
Peter B. Maloney, First Cavalry, California Volunteers, June 25, 1862.
Albert Schmidt, June 25, 1865.
Cassius A.B. Fisher, formerly of First Infantry, California Volunteers, February 19, 1867.
John Brownley, May 26, 1868.

Besides these graves there are fourteen more, whose head-boards bear the simple and pathetic inscription,

"UNKNOWN"

and eleven unmarked mounds. All these twenty-five were killed by Apaches. A bloody record, truly, for the little strip of territory almost within gunshot of the fort!

Nor must I omit mention of two more graves—those of Sergeant Robert Evans and private Jeremiah Lawson, of Troop C, Tenth Cavalry, who were killed by Indian *scouts* in Galeyville cañon,[4] January 3, 1886. Both men were Negroes—the Tenth is a black regiment—and it is believed that they abused

---

[3]Lieutenant Carroll and "a citizen" were killed together; the citizen is not named, but he might have been John Coss, mentioned below.

[4]This was Turkey Creek Canyon, according to Will C. Barnes, *Arizona Place Names,* 1st ed. (Tucson, University of Arizona, 1935).

the Indian scouts who were with their command. At all events, the scouts shot them both.

I have heretofore referred to

CAPT. JOHN G. BOURKE,

one of Gen. Crook's most trusted friends and officers. He is not only a polished gentleman, but in many respects a remarkable man. In the army for twenty-five years, and on the frontier for seventeen, he has used his opportunities well. He is of marked literary ability—none of your dude gushers, but a strong, terse, and often elegant, writer. With a wisdom which is so strange to me that so few men imitate, he has kept full and accurate notes of all the extremely interesting affairs that have come within his knowledge, and to-day has a whole library of note-books of great scientific and historical value.[5] He is now well known to scientific and ethnological institions in the East. I see, too, that the Scribners have brought out his little book, "An Apache Campaign,"[6] which is a vivid picture of some of the difficulties of hunting down these meteoric butchers. It is well worth reading. He has been ordered to Washington now to write up certain important notes. He is also compiling an Apache dictionary.

WHY CROOK QUIT.

I told you the other day that the change of commanders in the Department of Arizona, was made at Gen. Crook's request. The (official) Army and Navy Journal of April 10, adds to this a comment of its own. It says: "Gen. Crook's request to be relieved from the command of the Department of Arizona was telegraphed to the Lieutenant General [Philip Sheridan] about two weeks ago [it was before Geronimo's escape]. He assigned no reason for the desired change, but the impression is that he had become tired of the complaints from different quarters about his Indian policy, and thought he would retire for the purpose of letting the country see if any other officer could do better."

That is a pretty clever guess, I opine. Of course we all hope that

GEN. MILES

will be so fortunate as to wind up the outstanding renegades in short order; but if he does not, sensible men will not one whit lower their good opinion of him.

---

[5]Bourke's legacy of 128 closely-written and well-illustrated notebooks are in the West Point Library; they are available on microfilm.

[6]John G. Bourke, *An Apache Campaign in the Sierra Madre,* was published first in *Outing Magazine* in 1885, then as a book by Charles Scribner's Sons in 1886, and reprinted verbatim by that publishing company in 1958. The work was apparently written to refute criticisms of Crook's Sierra Madre expedition and events that transpired with relation to it. It therefore has polemical touches and some omissions, as noted in Thrapp, *General Crook and the Sierra Madre Adventure.* Yet Bourke's is an indispensable source work on that endeavor.

It is a question not only of skill and generalship, but also of good luck. Consider the vast territory which the hostiles range, nine-tenths of it absolute wildness, and think of catching thirty-four vigilant, tireless and fearless hostiles in it! If they were a thousand, the task would be comparatively easy, but as it is, the only way is to keep forever after them, and trust to luck to run across them sometime. To drop a needle off the Santa Rosa in Wilmington harbor and bet you could bring it up at the first dive, would be not a speculation but a sound business proposition beside catching these elusive refugees at the first clatter. Gen. Miles is moving—like the sensible man he evidently is—with wise discretion. He is not trying to be smarter than those who went before him, but is just gathering up the thousand intricate details of the complex situation, in a business-like way, and getting "a good ready." He is a man who puts on no frills, but is keen, practical, decisive and long-headed. You are apt to hear music from this direction, some of these fine mornings.

I had everything packed up this evening to start homeward on the morning train, planning to spend there the interim of preparation, and await telegraphic orders to seek the front again. That was on the supposition that several weeks would elapse before anything active was put on foot. But I had a long talk with Gen. Miles this evening, and though I am not at liberty now to tell what he said, the upshot of it all was that I will remain here. I have his promise that I shall accompany the first expedition after the hostiles. How's that for a bonanza?

LUM.

CHAPTER VII

Los Angeles Times, April 22, 1886:

GEN. CROOK.

A Review of His Indian Campaigns.

HARD SLEEPING OUT AND FIGHTING.

*How He Has Been Uncle Sam's Right-hand Man
for the Suppression of the Savages.*

[STAFF CORRESPONDENCE OF THE TIMES.]

FORT BOWIE, A.T., April 19, 1886.—By the time you read this I shall doubtless have jumped the picket-fence of civilization—if Arizona can be said to have any such boundary—and be plugging away over the rocky ranges of Chihuahua and Sonora. Everything is all fixed as fine as a little red wagon painted yellow, but unfortunately my tongue is tied as to all details whatever. There is one thing, however, which is dead certain and no secret—namely, that if my notes aren't written up p.d.q., I'm going to get left. It will be an extremely *dia fria* when we get any word to the outside world from that howling wilderness. So here goes for an installment.

From talks with Gen. Crook—who is unapproachable on nothing but his own achievements—and with others who have been able to put a little flesh upon the very bare skeletons he would give me, I have gleaned a very condensed sketch of the old hero's wonderful record. What a nation of ingrates we are! There is a man who has done more for the frontiers of the United States than any other man living—probably than any two men. For more than a generation he has been in the active military service of his country, and always successfully. But all this is lost sight of because in this little campaign he has captured only four-fifths of the hostiles. Bah! It makes me ashamed of my country! Let me refresh your memory a bit, as to what this cruelly persecuted man has done.

Born in the Buckeye State, Crook entered the Military Academy at West Point July 1, 1848, and graduated four years later with the rank of Brevet Second Lieutenant of the Fourth Infantry. July 7, 1853, he became a Second Lieutenant. In that year he entered upon the long series of Indian campaigns, in which he has been engaged ever since with the exception of his service in the civil war. His first work was in Northern-California and Oregon; and so distinguished were his skill, coolness and courage that the Legislature of California voted him a medal before he was a First Lieutenant.

THE HUMBOLDT INDIANS,

then on the war-path, engaged his initial attention, and he made short work of them. The Humboldts had a narrow range, and were "not much force." They were fish-eaters, poorly armed, and, says the modest general, "they never amounted to anything." His prompt and efficient handling of them, however, began the foundation of his fame as an Indian fighter.

Having wound up the Humboldts, he was sent to suppress

THE ROGUE RIVERS AND SHASTAS,

who, for years, had been harassing the mining regions. These were a hardier race than the Humboldts, inhabited a rougher and more impenetrable country, and were better armed. They had the muzzle loading fire-arms of that day, and, with sufficient numbers, would have proved a very formidable foe. But there were only a few of them, and they were soon crushed by the indomitable young lieutenant.

His next service was against the

PITT RIVERS, KLAMATHS AND TOLLAWALLAS[1]

in succession. Here he showed the same brilliant qualities which were already marking him as "a coming man," and March 11, 1855, he received a well-earned commission as a first-lieutenant.

In '58 Lieut. Crook went up to the Yakima country, under Major [Robert S.] Garnett[2] (later killed at Cheat River, on the Rebel side) to operate against the federation of

COLUMBIA RIVER TRIBES.

These were subjugated in three or four months, Crook winning new laurels in the short campaign.

He then went back to the Klamath, and was there until the breaking out of

THE CIVIL WAR,

which he entered as captain of the 4th Inf., May 14, 1861. Of his honorable career in the war, I need not go into detail, simply mentioning his commissions, as milestones in his progress. Sept. 17, 1861, he became colonel of the 36th Ohio Infantry; Sept. 7, 1962, brigadier-general (of volunteers); July 18

---

[1]Lummis probably means the Tolowa Indians of northern California, closely associated with Oregon Athapascan tribes.—John R. Swanton, *The Indian Tribes of North America* (Washington, Government Printing Office, 1953); A.L. Kroeber, *Handbook of the Indians of California* (Berkeley, California Book Company, Ltd., 1970).

[2]Garnett (1819–1861) was a West Point graduate from Virginia who served in artillery, cavalry and infantry branches, won two brevets in the Mexican War, and was killed as a Confederate Brigadier General at Carrick's Ford, Virginia.—Cullum.

1864, brevet major-general; October 21, 1864, major-general; July 18, 1866, major of the 3rd Infantry (regular); July 28, 1866, lieut.-colonel 23rd Inf. and honorably mustered out Jan. 15, 1866.

From the War of the Rebellion, Lieut. Col. Crook was called to conquer

THE PIUTES

in the far Northwest. He reached Boise City, Idaho, on the 11th of December, 1866. All through that big country the Indians had been a long-standing scourge and terror. "About a week after getting there," said the General, with a smile at the recollection, "I took the field with a command of 40 men, with my old clothes and a toothbrush, and didn't see a house again for two years." That army, 40 strong, must have seemed very imposing to an officer who had just come from a field where he commanded an army corps! But if there ever was a man who adapted himself to his surroundings, it was Crook. He marched out with his handful of men, and camped on the trail, sleeping out amid the snow and sagebrush until the last one of the hostiles surrendered, July 4, 1868. It was the first time that the Piutes had been at peace for many a long year. It was a terrible experience, that winter campaign in frozen Idaho. The cold was something awful. There was one time, when the mercury was at its lowest ebb, that the troops broke camp at midnight and marched, half frozen on their saddles, till dawn found them on the edge of the hostile camp. Crook drew up his cavalry and went swooping down on the village. His horse, an old campaigner, rose to the occasion; and taking the bit in his teeth, flashed far ahead of the command, and right into the midst of the astounded Indians. It put the rider, brave as he was, in a decidedly unpleasant predicament. He was between two fires, and the bullets from his own men whistled thicker around him than did those of the foe. As soon as possible, Crook dismounted and went to fighting on foot. His troops were now with him, and there was a short, fierce struggle among the wickiups. Crook and a red-headed soldier from Boise were making a flank movement together; and, in going around a wickiup, passed on either side of a tiny bush. There was a Piute behind that bush; and as they reached it, "Reddy" got a bullet through his heart, and Crook, jumping to one side, perforated the Piute.[3] The Piutes were a well-armed and warlike race, the strongest enemies (among Indians) against whom Crook had yet been pitted.

Having subjugated the Piutes, Crook was sent

BACK AFTER THE PITT RIVERS,

whom he had conquered eleven years before. His second tussle with them was

---

[3]In his autobiography Crook says the white casualty was a civilian from Silver City, Idaho, whose name he does not give.—*General George Crook: His Autobiography,* ed. by Martin F. Schmitt (Norman, University of Oklahoma Press, 1946), 149.

as successful as the first, and taught them a lasting lesson. He not only whipped them, but afterward applied the salve of peaceful teaching.

Thus far, the Indians with whom Crook had had to do were tribes of limited range, mostly in timbered country, and little given to forays of any extent. The Piutes, it is true, were considerable raiders, though nothing compared with the Apaches. Their country was a little like Arizona in many respects, but nothing like as rough.

THE APACHES

had been at their murderous tricks from time immemorial, and there was no doing anything with them. At last, in 1871, the now celebrated Crook was sent down from the north to try his hand at these indomitable savages. For a year he was so hampered by the corrupt "Indian ring" that he could take no active measures.[4] This time he used to good advantage in drilling a force of Apache scouts, putting in, to use his own words, "The hardest work I ever did." In mid-September, 1872, Crook got his hands loose and fell upon the hostile Apaches like a thunderbolt. It was a bloody struggle. The Apaches had not yet learned the art of running away, which is now their strongest hold. They stood their ground like men, and fought with savage skill. They had not yet achieved the breech-loader; and, in fact, most of them were armed only with bows and arrows. Hundreds of them were killed, and still the rest fought on. It was war upon the heels of war. The Apaches are not a family, but a nation, made up of a host of little tribes—the Chiricahuas, Tontos, Pinals, Aravipas, Hualapais, Mescaleros, Jicarillas, San Carloses, Warm Springs, Miembres, White Mountains, Coyoteros, Apache Yumas, Apache Mojaves (related to the western Yumas and Mojaves) and so on.[5] The Warm Spring and Miembres Apaches are subdivisions of the Chiricahuas. Campaign followed campaign, against tribe after tribe. An intelligent corporal, who served under Crook, said to me the other day:

"In the Tonto campaign in 1872, the General laid out in rain and snow and sleet for months, to get the Tontos, when no one else wanted to do it and no one else *would* do it. And he got them, after an awful campaign, in which he

---

[4]It was the Washington "peace effort" rather than the so-called Indian Ring that stayed Crook's hand until negotiators Vincent Coyler and O.O. Howard had a chance to bring in any peaceably-inclined Indians.

[5]Lummis here is a little shaky on his ethnology, accepting pioneer descriptions of virtually all militant Arizona Indians as "Apaches." Of his list, the true Apaches included the Chricahuas, Tontos, Pinals, Arivaipas (Aravipas), San Carlos, Warm Springs or Mimbres (Miembres), White Mountains or Coyoteros, Mescaleros and Jicarillas. The others (and he should have listed the Yavapais, against whom most of Crook's initial efforts were directed) were non-Apache peoples and in some cases distinctly hostile to the Apaches. The Jicarillas, of course, were Plains Apaches of New Mexico.

exposed himself as freely as his commonest soldier. And how he did thrash the Tontos! A lot of them fortified themselves in a cave, and he killed 76 of them right there.[6] And he pushed them to the wall that way, through the whole campaign. I don't believe the Chiricahuas would be on the warpath now if Crook had been let alone. He whipped the other tribes till they were ready to behave. But in 1876, just as he had the Chiricahuas cornered, and was about to paralyze them, Gen. O.O. Howard got a special dispensation from the President to treat with them and subdue them by peaceful measures.[7] He has said many harsh things of Crook, but the sharpest answer 'the old man' ever allowed himself to make was that 'in his experience with Indians he had never seen the wisdom of trying to subdue a ferocious tribe by the grace of God.' And the logic of events proved that he was right. You have to hammer those fellows till they're *afraid* to go on the warpath. Crook was never allowed to do this to the Chiricahuas. Howard put them unpunished upon the reservation, and they soon stampeded into Sonora, carrying out their usual programme of butchery. In '83 they broke out again; and now they are raiding still. If Crook had been let at them when he had them within reach, they would be as quiet to-day as the tribes which Crook did whip into peace."

The speaker is a man who has been in a position to know whereof he affirms, and he hits the nail pretty squarely.

Well, Crook made a glorious record. He fought the Apaches from the Grand Canyon of the Colorado down to Fort Grant (28 miles north of Willcox), and from the Little Colorado to the road between Tucson and Yuma. And at last, worn out, crushed and humble, the Apaches came to

THEIR APPOMATTOX.

On the 6th of April, 1874,[8] they surrendered—5,000 of them—to the Gray Fox at Verde. It broke the backbone of the Apache nation, and since that time their outbreaks have been merely sporadic little eruptions of a tribe or a small gang of malcontents. It won a star for the matchless Indian fighter who had accomplished such a wonderful result; and on the 29th of October, 1873,

---

[6]This famous attack at "Skeleton Cave" on the Salt River involved Yavapais in all probability rather than Tonto Apaches. Crook himself did not take part in the action, the commander being Captain William H. Brown.

[7]Howard's memorable visit with Cochise and his agreement with that Indian occurred in 1872, not 1876. Whether Crook could have dealt with the Chiricahuas better at that time is a matter for speculation. Cochise held his people at peace with the Americans—though not with the Mexicans—from 1872 until his death in 1874. After concentration on San Carlos, which occurred in 1876, Chiricahua outbreaks were common until Crook brought the Indians back from Sonora in 1883, and even until the Geronimo bust-out of 1885.

[8]He means 1873.

Lieut. Colonel George Crook was made Brigadier General Crook. Never was that high position better earned.

It was with the utmost difficulty that I extracted any information about Crook's achievements from himself. He has little use for newspaper men; and it was only through the fact that I brought very kind credentials from one of his old comrades in arms[9] that I could approach him on such subjects at all. He is not starchy in any respect, and talked freely and animatedly about hunting, fishing and other topics in which we could meet on common ground. But when it came to what he had done, and to the influences which have fought him in his operations, he closed his lips at once. The dates and similar outlines of his campaigns he told me; and now and then, warming at some old recollection, a humorous incident in his experience. Speaking of the big surrender above noted, reminded him of

CHIEF DELCHÉ,

then one of the Apache leaders. Delché[10] was not the George Washington of his tribe. In fact, his name means, in Apache, "the liar." He used to come in to the agencies and make peace, picking up a pebble and saying, "When this stone melts, I will change." Then he would turn right around, murder somebody, and be off again on the warpath. But when Crook got after him, Delché amassed a belly-full. At the surrender he came up to Crook and said: "Last fall I had 120 warriors, and thought I could whip the world. Now I have only 25. Your copper cartridges have done it. We are all worn out. We are nothing but skin and bones, and can't sleep nights. A coyote starts a stone rolling down hill, and we think it is the soldiers coming, and we run. We are tired of running, and want to be at peace." Old Shuttlepan,[11] another raiding chief, said: "I surrender, not because I love you, but because I fear you"—and the savage scowl on his face emphasized the honesty of his words.

"TO SHOW THEIR CHARACTER,"

said Crook, "take Delché's case. After he was so glad to surrender we put him on the reservation at Verde. After a while he got uneasy and surrounded the agent, who would have been killed right there if it hadn't been for some of our

[9]This no doubt is a reference to brevet Lieutenant Colonel Harrison Gray Otis (he would become brevet Major General in the Spanish-American War), at the time largely owner of the *Los Angeles Times* and its editor-in-chief.

[10]Delshay (c. 1835–1874) was either a Tonto Apache or a Yavapais. Because he was an able leader of his people, he was most troublesome to the whites, but he was as much sinned against as sinning.—Thrapp, *The Conquest of Apacheria,* 140 ff.

[11]Eschetlepan after his surrender served several enlistments as a scout under Crook's policy of persuading the wildest of the former hostiles to scout for other Indians still out.

[12]This incident is described more fully in Dan L. Thrapp, *Al Sieber, Chief of Scouts* (Norman, University of Oklahoma Press, 1964), 121–24.

Apache scouts.[12] That night Delché and his whole gang broke out again. They got into the river and rode in it 20 miles, often having to swim their horses. This was to hide their trail. They kept on in the water until they reached a point where a long ledge of rocks runs down into the river, and, riding out on this without leaving a sign of a trail for some miles, off they went. About the same time some of the Apaches, down here broke out and killed several people. We went after them, chased them down, killed some, and the rest came back. I broke their arrows and smashed their guns. They said they would live quietly on the reservation; but I wouldn't believe them, and told them they'd have to go out again and that we would follow them up and fight them. They begged and begged; and at last I told them that if they would

"BRING ME THE HEADS

of the seven ringleaders. I would given them another chance. Sure enough, they brought me the heads. {I understand from another source that the General was on his porch at head quarters when some of the penitents came up, and lifting a burlap sack, poured out the seven ghastly trophies before his astonished eyes.—L}. Then I went up to Verde and sent for Delché's head. They brought it to me soon. Then, when I got back to San Carlos, somebody else brought me Delché's head! I don't know which was Delché's own head—both looked like him—but it is certain that they got all he had, for he never turned up again!"

But Crook's work with the Apaches had only its preface in the surrender of April, '73. The conquered tribes were put upon reservations throughout the Territory and at last were consolidated upon the White Mountain Reservation. Their conqueror, as great in Indian control as in Indian warfare, bent all his tireless energy and unparalleled experience, to making good Indians of them. He taught them, with fatherly patience, the arts and the value of peace and agriculture. I shall have more to say of this hereafter.

It had come to be understood throughout the country that when there was a difficult Indian problem on hand, George Crook was the man to solve it. And so it came to pass that when the

SIOUX AND THEIR ALLIES,

the Cheyennes and Arapahoes, baffled all others, Crook was sent up to Wyoming and Dakota, in 1875, to subdue them. There were 15,000 of these Indians—among the most noted savage warriors in the world. They were different from any other Indians with whom Crook had had to do, being a race of centaurs—the most expert horsemen alive, probably. Like other Indians of the plains, they were no good afoot; but mounted on their hardy ponies they were the scourge of their broad range. They had several thousand head of horses; and when the warrior had tired one mount he sprang upon another. During the long winters they were holed up in comfortable quarters. Their ponies got too weak to be of any use for a long time, and they would not start out on a raid till the grass was well up in the spring. These savage warriors had had everything their own way, hitherto. After the famous Fetterman massacre,[13] 10 years before, they had dictated their own terms. They even made the government evacuate three big forts. Crook settled the Sioux problem. A week after the brave but reckless Custer had thrown away his own life and the lives of his whole command—four companies—by dashing into 4000 or 6000 warriors, Crook, with only 100 more men than Custer had, met the hostiles, outgeneraled and routed them.[14] He was using the same subtle means of disintegrating them that he had employed down here—using scouts of their own blood against them. I wouldn't rob Gen. Miles of one of his well-earned laurels; but, when you talk of Sitting Bull, do not forget old Spotted Tail, the real head of the Sioux. Sitting Bull was like Geronimo and some of our own newspaper generals of the Rebellion—bigger in type than in the field. Crook sent Spotted Tail away, and made his followers surrender.[15] In 1877, at President Hayes's request, he moved that great body of Indians from a point 100 miles north of the Northern Pacific down to the Missouri River, a journey of 300 or 400 miles, through the rigors of a terrific winter. From that time to 1882 he was in the north, attending to these tribes. Of his subsequent operations in Arizona and Mexico I must defer the telling till another letter.

LUM.

[13]Captain William J. Fetterman and more than eighty persons under his command were wiped out to a man by Sioux north of Fort Phil Kearny, Wyoming, on Dec. 21, 1866. The Indian victory was the most celebrated in the West until the Custer fight a decade later.

[14]Historians would dispute Lummis's evaluation of Crook's action on the Rosebud, which occurred on June 17—before, not after, Custer's defeat June 25, 1876. Most students consider it at best a stand-off; some believe it to have been an outright defeat for Crook.

[15]Spotted Tail was chief of the Brulé Sioux. He remained at peace and far from the theatre of conflict when Sitting Bull, a Hunkpapa Teton medicine man, the Oglala chief Crazy Horse, and others made war against Custer and Crook. Spotted Tail generally was peaceably inclined and caused little trouble.—George E. Hyde, *Spotted Tail's Folk: A History of the Brulé Sioux* (Norman, University of Oklahoma Press, 1961), 220–33.

CHAPTER VIII

Los Angeles Times, April 25, 1886:

A REMINISCENCE

Of the Indian Campaign of 1883.

THE BLOODY RAID OF GERONIMO'S BAND

*How Juh Lost His Life—Slippery Apaches—General Crook's
Preparations to Cross the Mexican Border.*

[SPECIAL CORRESPONDENCE OF THE TIMES.]

FORT BOWIE, A.T., April 20, 1886.—In September, 1881, the famous Juh
(pronounced Hoo) with Geronimo, then high right hand man, and their bands,
went down into Mexico and apparently settled there. Juh lost his life near
Casas Grandes, Chihuahua, and thus one of the most dangerous of Apaches
was removed. He was very drunk, one day, and as he rode along a high bluff
over the river, his mule slipped and they both went down to death. Juh was a
hard, merciless savage, of great determination, and his name was a terror all
along the border. He was 36 or 37 years old, nearly six feet tall in his moccasins, very dark skinned, and with an impediment in his speech. American
troops were not at that time allowed to cross the line, and the Chiricahuas
were secure in Mexico. In the spring of '82, a band of them came up to this
country, and forced Loco, a chief of the Warm Spring Apaches, to go out
with them. American troops followed them down below the border, but
were met at the northern end of the Sierra Madres by the Mexican General
Garcia,[1] who ordered them back.

EVERY WARM SPRING AND CHIRICAHUA

Indian from the reservation was now in Mexico, except Toklonnay,[2] who
remained faithful to the Americans from first to last. When Crook came back

---

[1]The Mexican officer was Colonel Lorenzo García, in command of a body of regulars which
ambushed the southward-moving Apaches and killed about seventy-five of them, including
eleven warriors; the others escaped into the Sierra Madre. For a biographical treatment of Juh, a
remarkable southern Chiricahua chief, see Dan L. Thrapp, *Juh: An Incredible Indian* (El Paso,
Texas Western Press, 1973). Juh was about sixty at his death, which may not have resulted from
drunkenness, as the whites liked to persuade themselves.

[2]Toklanny, who took the first name of Roger, was born about 1863 and died in 1947. He was
a Mimbres, said to have served longer as a U.S. scout than any other Apache, and he never bore
arms against the American whites. Yet he was sent as a prisoner with the Chiricahuas-Mimbres
to Florida and was held in that nominal status until permitted to return to the Mescalero
Reservation in 1913. He died in New Mexico.—Griswold.

(Sept. 4, '82) from his hard work among the Sioux, this is the condition of affairs he found—half a thousand of the Apaches dwelling in impregnable strongholds far below the Mexican lines. They showed no disposition to come north; but Crook, with his wonderful knowledge of their character, felt certain that it was only a question of time when they would make one of their bloody raids across our border. Foreseeing this, he did everything in his power

TO DIVERT IT.

Early in October, '82, he went down to the Mexican line and tried to open communications with the renegades, but found it impossible. Failing in this, he pushed the reorganization of the pack-trains of the department, which he had begun immediately upon resuming command, and posted his troops along the frontier at the points whence they could best spring to instant service, should the renegades bob up above the line. He also put Capt. Crawford and a body of Apache scouts near Cloverdale, N.M., (once notorious as headquarters for rustlers) to patrol the line from there westward. Some of Crawford's scouts sneaked down below Casas Grandes, Chihuahua, in search of the renegades, but could find no trace of them. They had retired into the deeper recesses of the Sierra Madres.

THE CAMPAIGN OF 1883.

You have heard more or less, from unofficial sources, of the remarkable Apache campaign of 1883; but, so far as I am aware, these accounts were more or less inaccurate, and a brief account of the operations, gleaned from official reports, may prove of interest.

From the date (September 4, '82) of Crook's return to the Department of Arizona up to the latter part of March, '83, there was not a depredation or outrage of any sort committed in Arizona by any Apache, either of those on the reservation of [or] the renegades. This period was improved by Crook, not only in the preparations—already meagerly noted—in anticipation of a raid from below the border, but also in

CONSCIENTIOUS WORK

among the Indians on the reservation. He impressed upon them the necessity of becoming civilized and self-sustaining; and, as an essential to the latter, told them they could select suitable homes anywhere inside the reservation, instead of roving nomadically over the whole of it. The head men of the respective bands were to be held responsible for the behavior of their people thus scattered. Some of the brightest, best and most influential of them would be enlisted as soldiers, but would reside among the people, and would assist in leading them toward self-government. Whenever a tribe showed its incapacity of self-control, it would be brought in to the agency where it could be controlled. If any band became bad, the other bands must join together and control it. He promised to bring in the white soldiers only when the Apaches proved themselves incapable of self-control. One condition of their being allowed to pick out individual homes on the reservation, was that they must support themselves after their crops came in in the fall. The chief would be held responsible if any tizwin was made. They must put their spare money into horses and cattle. Their future would depend upon themselves.

CHATTO'S RAID.

At last the event foreseen by Crook came to pass. Early in March, 1883, the renegades became weary of living idle in the Sierra Madres. A party of about 50 under Geronimo raided into Sonora after stock; while Chatto,[3] one of the most energetic of the Chiricahuas, took 25 companions to raid Arizona after a fresh supply of ammunition. Chatto and his band crossed the national line, near the Huachuca mountains, March 21, 1883. At sunset the same evening, they killed four white men at a charcoal camp, 12 miles southwest of Fort Huachuca. One of the raiders was killed in the fight. On the following afternoon they massacred three more men near the Total Wreck mine, on the east side of the Whetstone mountains; and the same night crossed the S.P.R.R. near Benson. On the 23d, two more men fell victims to the raiders near the south end of the Galiuro mountains.

From here the raiders broke up into small parties, their trails leading across the Pinaleno mountains—the northern extension of the Chiricahua range—across the San Simon valley, and through the Peloncillo mountains to the Gila valley near Ash Springs. All these marches, of course, were made in the night. The raiders crossed into New Mexico; and, on the 28th, butchered Judge [H.C.] McComas and wife, on the stage road between Silver City and Lordsburg; and two men on the Gila.

As soon as it became known that the renegades had raided from Mexico into Arizona, a general and

VIGOROUS PURSUIT

was begun by Crook's forces. But so like lightning was the sweep of the hostiles that pursuit was vain. Their speed is shown by the fact that they were in

---

[3]Chatto (1854–1934) was one of the more famed Apache raiders, although some believe that Chihuahua, rather than Chatto, led the 1883 raid generally credited to him. Its most-remembered incident was the slaying of Judge and Mrs. H.C. McComas and the kidnapping of their son in southwestern New Mexico. After Chatto came in, following Crook's Sierra Madre expedition, he became reconciled with the whites and served in a valuable way as a scout for Britton Davis and other officers, although he was sent into exile with his people. He died of injuries suffered in an automobile accident on the Mescalero Reservation—Griswold; Frank C. Lockwood, *The Apache Indians* (New York, The Macmillan Co., 1938).

Arizona only six days at the outside; they traveled nearly 400 miles through a rough country. Indeed, they had to hurry so that they failed to secure the ammunition which was the object of their raid.

Crook had foreseen, as soon as the news of their presence in the Territory was wired him, that it would in all probability be impossible to catch the raiders by direct pursuit; and while continuing the pursuit with the utmost vigor, he made every possible preparation to

INTERCEPT THEIR RETURN

to Mexico. He instantly telegraphed to the commanding officer at Fort Bowie to scout the Chiricahua mountains and patrol the San Simon and Sulphur Springs valleys; to Fort Thomas, to send two companies to Nogales; to Fort Grant, to send two companies to White River; to Fort Huachuca, to keep the country between the Huachuca and Dragoon mountains thoroughly scouted; to Capt. Crawford, Cloverdale, to put his scouts in position to intercept the raiders if they tried to return through the Stein's Peak range, or the Los Animas plains; to Lieut. Britton Davis,[4] at San Carlos, to watch closely for them at the reservation. He also, by telegraph, ordered Lieut. [Charles B.] Gatewood's[5] scouts to Huachuca: all the cavalry at Fort McDowell, and four cavalry troops from Fort Apache to Willcox. All these orders were promptly and conscientiously carried out; the pursuing parties rode their best—and yet not a soul saw one of the raiders! Crook had fenced the boundary line as well as he could; but troops can be put only where there is water—and in this awful land it is, as one gubernatorial southerner remarked to another g.s. on a celebrated occasion, "a

---

[4]Davis (1860–1930) was born in Texas, the son of a Union supporter and later stormy governor of the postwar state. He was graduated from West Point in 1881, joining the 3rd Cavalry. Posted to Arizona, he was named to command Indian scouts, where his intelligent and intrepid service won the commendation of Crook. After a grueling scout into the Sierra Madre of Sonora chasing Geronimo in 1885, Davis resigned. He became the manager of extensive Chihuahua mining and cattle interests, accumulated wealth, but lost virtually everything in the Mexican Revolution. He then settled at Congers, New York, and later at San Diego, California, where he wrote his book and where he died.—Davis, *The Truth About Geronimo,* Lakeside edition (Chicago, R.R. Donnelley & Sons Co., 1951).

[5]Gatewood (1853–1896) was one of the true heroes of the Apache wars. Born in Virginia of a Confederate-inclined family, he was graduated from West Point and assigned to the 6th Cavalry in 1877. He saw more extensive Apache experience than perhaps any other officer, beginning with numerous actions of the Victorio war of 1879–80 in New Mexico and Old Mexico. He went into the Sierra Madre in 1883 with Crook, commanding Indian scouts; during the Geronimo campaign of 1885–86 his services were extensive and, in fact, indispensable. To Gatewood belongs the credit for finally bringing in the last of the hostiles, although he received none of the four Medals of Honor given during the campaign and was otherwise professionally slighted. He was badly injured in the Johnson County War of 1892 in Wyoming and was retired for disability. That he was never in his lifetime given the recognition which was his due is to the discredit of his country.

long time between drinks." Easily as the swallow darts through between the poles of a telegraph line, the elusive raiders slipped through the necessarily coarse meshes Crook had spread for them; and sweeping down by night through the mountains on the east side of the Los Animas valley, N.M., were safely

BACK IN THEIR STRONGHOLDS.

Crook had telegraphed to Washington for instructions, and on the evening of March 31 received the following satisfactory reply by telegraph:

PRESIDIO OF S.F., March 31, 1883.

Commanding General Department of Arizona, Sir: Instructions, just received from the General of the Army, authorize you, under existing orders, to destroy hostile Apaches, to pursue them regardless of department or national lines, and to proceed to such points as you deem advisable. He adds that Gen. [Ranald S.] Mackenzie's forces [in New Mexico] will co-operate to the fullest extent.

By order of General [John M.] Schofield.

(signed) [John C.] Kelton, A.A.G.

Crook immediately hurried by rail to Guaymas and Hermosillo, in Sonora, and to the city of Chihuahua in the State of the same name to

CONSULT THE MEXICAN AUTHORITIES,

civil and military, as to an amicable agreement for "carrying the war into Africa." He says: "The reception extended me was of the most hospitable and cordial character. Gens. [José Guillermo] Carbo[6] and [Bonifacio] Topete[7] and their staffs, in Sonora, and Gov. [Luis] Torres and other prominent functionaries in that State; Govs. [Mariano] Samaniego and Terrasses,[8] of the State of Chihuahua; Mayor [Juan] Zubirán, of the city of Chihuahua, and other gentlemen—all gave assurance that they would, in every possible way, aid in the subjugation of the Chiricahuas, who had for so many years, murdered and plundered their people, as well as our own. Consuls [Alexander] Willard, at Guaymas, and [Louis H.] Scott, at Chihuahua, rendered me valuable assistance."

These important details arranged, Crook started for San Bernardino

[6]Brigadier General Carbo commanded the First Military Zone of Mexico, which included Sonora and certain other states.

[7]Brigadier General Topete, of Hermosillo, served under Carbo and was in charge of immediate operations against Apaches in Sonora.

[8]It is not certain whether Crook conferred with Luis Terrazas, a former governor of Chihuahua and a political and economic power of the state, or with Colonel Joaquin Terrazas, his cousin, who had destroyed Victorio three years earlier.

Springs, on the Mexican line, arriving there April 29, with his forces. Before burying himself in the Sierra Madres, it was necessary to protect the head and flanks of his expedition, and to protect the settlers of Arizona from raids during his absence. For this double purpose he made the following

DISPOSITIONS OF TROOPS:

Maj. James Biddle[9] was left with five companies of the Third and Sixth Cavalry, at Silver creek; Capt. G.E. Overton,[10] with two companies of the Sixth Cavalry, at old Camp Rucker; Capt. P.D. Vroom,[11] with two companies, Third Cavalry, at Calabasas. These troops, in conjunction with those from Fort Bowie, under Capt. [William A.] Rafferty,[12] and from Fort Huachuca, under Maj. [Nicholas] Nolan,[13] were to patrol the country with all possible thoroughness. Col. E. A. Carr[14] Sixth Cavalry, was authorized to assume general command of these forces, should it at any time become necessary. Capt. Wm. E. Dougherty,[15] commanding officer at Fort Apache, was sent back to his post to attend to the control of the Apaches on the reservation.

Everything being now prepared, Crook

OPENED THE CAMPAIGN,

May 1, leaving San Bernardino Springs with the following force: 193 Apache

---

[9]Born in Pennsylvania, Biddle (1832–1910) was commissioned from New York in 1861 and had a good Civil War record. He joined the 6th Cavalry in 1870. Biddle was colonel of the 9th Cavalry at his retirement in 1896, and in 1904 he was made a Brigadier General.—Heitman.

[10]A New Yorker, Gilbert Overton (1845–1907) was commissioned in a state cavalry outfit in 1861 and after the Civil War joined the 6th Cavalry. He won a brevet for gallantry against the Comanche Indians. Overton retired in 1891.—Information from Lieutenant Colonel David Perry Perrine.

[11]Peter Vroom (1842–1926), of the 3rd Cavalry at this time, was born in New Jersey, the son of a governor of that state, and retired in 1903 a Brigadier General.—Heitman; *Who Was Who.*

[12]Rafferty (1842–1902) was a West Pointer from New Jersey. He was co-founder of Fort Huachuca and saw much service against Indians, winning a brevet for an 1870 Comanche fight in Texas and serving too against the Cheyennes. He was in a hard Apache fight at Sierra Enmedio, Sonora, in 1882. He died from injuries suffered in an accident in the Philippines.—Heitman; Thrapp, *Dictionary of Frontier Characters.*

[13]Born in Ireland, Nolan enlisted in the army in 1852 and won a field commission ten years later, rising to become major of the 3rd Cavalry by 1882. He died the next year.—Heitman.

[14]Carr (1830–1910) was a West Pointer from New York who became a Brigadier General of Volunteers during the Civil War and a Brigadier General in the regular Army in 1892. He won a Medal of Honor at Pea Ridge, Arkansas, in 1862, and held important commands during his postwar Indian fighting days. He commanded the column that was involved in the Cibecue,

scouts, under Capt. Emmet Crawford, 3rd Cav., Lieut. C.B. Gatewood, 6th Cav., and Lieut. J.O. Mackay,[16] 3rd Cav., 42 enlisted men of the 6th Cav., under Capt. A.R. Chaffee[17] and Lieutenants Frank West[18] and W.W. Forsyth.[19] Acting Assistant Surgeon George Andrews and Hospital Steward J. B. Sweeney accompanied the expedition. Crook's personal staff consisted of Capt. John G. Bourke, A.A.A.G., and [First] Lieut. [Gustav Joseph] Fiebeger,[20] A.A.D.C. "This," says Crook, "was the maximum force which could be supplied by the use of every available pack animal in the department (over 350 animals; in excellent trim); and the minimum with which I could hope to be successful in the undertaking upon which I had engaged. We had supplies, field rations, for sixty days, and 150 rounds of ammunition to the man. To reduce baggage, officers and men carried only such clothing and

Arizona, fight which launched the long series of incidents culminating in Crook's 1883 expedition into Sonora.—James T. King, *War Eagle: A Life of General Eugene A. Carr* (Lincoln, University of Nebraska Press, 1963).

[15]Irish-born Dougherty (1841–1915) enlisted in the Civil War and was commissioned in 1863. He served on the Indian frontier from 1874 until 1883. Dougherty saw service in Cuba in 1898 and the Philippines in 1901–1902, becoming a Brigadier General in 1904, the year in which he retired. He lived in California until his death.

[16]Nova Scotia-born Mackay (1857–1911) was a West Point graduate from Nevada who became captain of the 3rd Cavalry in 1891 and retired nine years later.—Heitman; Cullum.

[17]Adna Chaffee (1842–1914) was a distinguished American soldier whose capacities were greater than the opportunities in the Army of his day. He was born at Orwell, Ohio, enlisted in the 6th Cavalry at the outbreak of the Civil War, and remained with the regiment for twenty-seven years. He was commissioned in 1863. In addition to his Arizona experience, he participated in the Cuban war of 1898, commanded the American relief expedition in the Boxer uprising in China, commanded in the Philippines for two years, was made Lieutenant General and Chief of Staff for the Army, and retired to Los Angeles in 1906.—William Harding Carter, *The Life of Lieutenant General Chaffee* (Chicago, University of Chicago Press, 1917).

[18]West (1850–1923) went to West Point and joined the 6th Cavalry in 1872, being active in South Plains Indian campaigns before reaching Arizona in 1875. He won a Medal of Honor at Big Dry Wash, Arizona, in 1882 and after leaving the Territory took part in the Wounded Knee Sioux operation of 1890–91 and helped police the Johnson County, Wyoming, feuding area in 1892. He retired in 1914.—Heitman; Thrapp, *Dictionary of Frontier Characters.*

[19]Forsyth (1856–1917) was a West Pointer who accompanied Crook into the Sierra Madre and did good work in the Geronimo war of 1885–86; he was in on the surrender of Geronimo in the latter year. Afterward, Forsyth took part in the China Relief Expedition of 1900 and was superintendent of Yellowstone National Park for four years. He died in Virginia.—Thrapp, *Dictionary of Frontier Characters.*

[20]Fiebeger (1858–1939) was an engineers officer who was taken along with the Crook expedition to collect notes from which the expedition's course could be accurately mapped; this was done. He was a West Point graduate who had a distinguished career, taught civil engineering from time to time, and died at Washington, D.C.—Thrapp, *General Crook and the Sierra Madre Adventure.*

bedding as was absolutely necessary and instead of keeping up their own messes, the officers shared the food of the packers."

On the 27th of March, while the hostiles were raiding up here, one of their number had deserted them, was arrested by Lieut. Davis [at San Carlos], and turned over to Crook. Pe-nal-tishn, commonly nicknamed "Peaches,"[21] after a severe cross-examination, had agreed to lead Crook to the hostile stronghold; and under the guidance of this Chiricahua, the expedition started out into the savage and unknown fastness of the Sierra Madres.[22]

LUM.

Los Angeles Times, April 29, 1886:

THE SIERRA MADRE

Conclusion of Crook's Campaign of 1883.

A MOST REMARKABLE ACHIEVEMENT

Over 500 Hostiles Captured, Without the Loss of a Man—The Way to Deal With Indians.

Crossing the line into Mexico, the expedition of 1883 followed down the San Bernardino river, the northernmost branch of the Yaqui. For three days they did not encounter a soul. The whole region through which they were passing had been depopulated by the Apaches; and great areas of former farms had become a jungly waste of mesquite and canebrake. They followed the hostile trail deeper and deeper into the wilderness, conducted by Pe-nal-tishn,

---

[21]Tsoe, or Pan-al-tishn, nicknamed Peaches because of his complexion, became the key to the success of Crook's great adventure. Peaches may have deserted the hostiles because he was not himself a Chiricahua, although he was married to one. He was a member of the Canyon Creek band, Cibecue group, of White Mountain Apaches. Because he was not Chiricahua, he was not sent into exile with that people, but probably lived out his life on the present Fort Apache Reservation where Wharfield, in 1961, met one of his sons.—Keith Basso, ed., *Western Apache Raiding and Warfare: From the Notes of Grenville Goodwin* (Tucson, University of Arizona Press, 1971); H.B. Wharfield, *Cooley* (El Cajon, Calif., pp. 1966).

[22]See Thrapp, *General Crook and the Sierra Madre Adventure* for an analysis of this singular military adventure, whose importance has often been overlooked by present-day historians. The work traces antecedents of the undertaking back to the affair at Cibecue, explores the succeeding incidents building to the knotty situation Crook proposed by this extraordinary means to unravel, and relates the adventure itself on the basis of primary accounts of participants and others.

or "Peaches," the Chiricahua who had deserted from Chatto's command on the 28th of March. On the 6th of May the expedition passed the little Sonorian hamlets of Bavispe, San Miguel and Basaraca, whose inhabitants were wild with joy.[23] They had good cause to welcome the coming of Crook. Their condition was deplorable. In all these little villages the swoop of the Apache was to be looked for at any moment. The people were absolutely cowed, and no man dared venture a couple of miles from home. It was such a reign of terror as obtains among the hamlets of India, where the man-eater prowls in the neighboring jungle. The people of Bavispe offered Crook the assistance of all their able-bodied men; but he declined for want of transportation and supplies. He also declined their proffer of four guides to the Sierra Madre, knowing that the Apache trailers whom he already had were matchless.

Moving cautiously and by night, to escape detection, Crook's forces

ENTERED THE SIERRA MADRES

on the 8th of May. It was evident that they were approaching the stronghold of the hostiles. Indian "sign" became abundant, and the invaders came upon abandoned camps which had been occupied by fifteen, twenty, thirty and even forty families; and upon cattle, horses and burros, alive and dead.

The country was indescribably rugged—lofty and inaccessible mountains, dizzy canyons and frowning cliffs. It was here that the expedition encountered the difficulties I referred to in my last. A good many mules were killed or crippled by tumbling down the terrific precipices; and traveling was tremendously arduous and slow. It was necessary, also, to use the utmost caution in proceeding, to guard against ambushes and against alarming the wary foe. There were, however, advantages not usually found in Apache-hunting— abundance of pure water and good fuel, the mountains being covered with oak and pine and full of running streams.

On the 12th of May, "Peaches" led the command to

THE ENEMY'S STRONGHOLD,

a fortified point far up among the rocks, and absolutely impregnable to attack. "For that matter," says Crook—and so say all who have been in that section— "the whole Sierra Madre is a natural fortress; and to drive the Chiricahuas from

---

[23]After following the course of the San Bernardino River from the meeting point of Arizona–New Mexico and Sonora southward, they had come to the Bavispe River. They turned up this stream, moving in a southeasterly direction, until they came to the historic communities Lummis mentions.

it by any other method, than those which we employed would have cost hundreds of lives." Reaching the stronghold, they found that the enemy, Apache-like, had already evacuated. Bred to a warfare which is made up of surprises, the Apache never stays in the same camp more than a few days at the outside. By continual shifting, he avoids much of the probability of being caught unawares. He never goes into camp, even for a night, when on the warpath, without securely fortifying himself; but no place can be sufficiently fortified to satisfy him as a permanent fortress.

But, though the hostiles had moved on, it was evident that they were not far away. So Crook left the pack-trains in the stronghold, under guard of Captain Chaffee's company, and sent out Crawford and the Apache scouts to scour the country thoroughly in front and on both flanks.

On the 15th of May, the scouts

STRUCK THE HOSTILE CAMPS,

which proved to be occupied by the forces of Chatto and Bonito.[24] Crook had given careful directions for surrounding these camps, but some of the scouts imprudently fired on a buck and squaw whom they espied, and a general fight ensued, lasting several hours. The hostiles were completely routed, nine or more being killed, while five half grown girls and boys were captured along with all the contents of the camps. Among the captured stuff was much that had been stolen by the raiders from Mexicans and Americans, including 40 horses and mules.

The hostiles were now so thoroughly alarmed that to pursue them was out of the question. They had scattered among the countless rugged peaks, whose every rock was a fortress from behind which they could deal back death with their breech loading rifles. There were but two alternatives. One was to go back to the United States, wait until the Chiricahuas felt secure and attempt to surprise them again. The other was to get them to surrender. The eldest of the girls captured in Chatto's camp said that if they would let her go she would call in some of her people for a pow-wow. She appeared sincere, and Crook allowed her to go. Next day, May 17, the scouts made a signal smoke, and six squaws came in. Crook refused to parley with them, telling them that he would talk

[24]Bonito probably was a Chiricahua who had a small following of four or five warriors, although he may have been a Mimbres or eastern Chiricahua. An Indian of that name figured in an affair at Fort Tularosa, New Mexico, in 1873, and one of his name was listed in an 1876 census at a Mimbres Reserve. At any rate, Bonito and his band bolted from San Carlos after the Cibecue incident and gained Old Mexico. He accompanied Crook back to the Reservation. If he survived and went east with the Chiricahuas in 1886, he must have succumbed before the people reached Fort Sill, for he is not mentioned there.

business only with the representative men of their tribe, whom they went out to fetch. Early on the 15th,

CHIHUAHUA CAME IN

and had a long talk with Crook. He said they had supposed the Sierra Madres to be impregnable—none of their enemies had ever penetrated their fastnesses before. The Mexican troops never got beyond the foothills. The Chiricahuas didn't waste bullets on the Mexicans, but simply rolled rocks down on them. He expressed the universal Apache hatred for the Mexicans, saying that the Mexicans took care to kill the Chiricahua women and children, but to run away from the men, and to treat them all with treachery. Shortly before this conversation the Mexicans had invited a lot of Chiricahuas into a small town near Casas Grandes, with every show of hospitality, had got them drunk and then murdered a lot, making prisoners of the rest. Chihuahua said the Chiricahuas would be glad to settle down in peace—they were tired of this incessant war.

After Chihuahua, the rest of the Chiricahuas came flocking in from all points of the compass.

ALL SURRENDERED.

Among them were the chiefs Geronimo, Chatto, Bonito, Loco,[25] Nachita and Keowtennay. The latter, of whom I shall have considerable to say hereafter, had never been on the reservation, having been born and reared in the Sierra Madre. The surrender brought into Crook's hands over 500 hostiles, including 120 bucks. They wanted to make peace and go back upon the Reservation. Crook told them no, he was tired of their disobedience and outbreaks. The best thing they could do was to stay right there and he'd stay and fight it out with them. He kept them on this ragged edge for several days; and each day they became more worried and more importunate, till at last they fairly begged him to take them to San Carlos. Geronimo and the others said, "we give ourselves up, do with us as you please."

Rations were now running short, and with this great number of prisoners on hand, it was all Crook could do to get back to his base of supplies without starving. He started homeward at once, therefore, the Chiricahuas sending out

---

[25]Loco (1823–1905) was one of the famous Chiricahuas, notable for his determined leaning toward peace with the whites rather than war, and for the high regard in which he was held by Crook and many other ranking Anglos. He and Victorio became co-chiefs of the Mimbres after the deaths of Mangas Coloradas and Delgadito. In 1885 he refused to go out with Geronimo, remaining with his band at peace on San Carlos; nevertheless, he was bundled into exile with the erstwhile hostiles. He died at Fort Sill.—John A. Shapard Jr., biography of Loco in preparation.

their runners to call in their stragglers who were scattered all over that frightfully broken country. Crook marched north, accompanied by such of the prisoners as had come in; while all the rest came straggling along behind as they received the news. *Every one came to the Reservation.*

NOT DISARMED.

Crook did not disarm any of his prisoners at the surrender. He has been damned so widely for his habit in this respect, by those who know nothing of the subject upon which they are so fond of exploiting their mouths, that I append the excellent reasons he gives in his official report:

"It is not advisable to let an Indian think you are afraid of him even when fully armed. Show him that at his best he is no match for you. It is not practicable to disarm Indians. Their arms can never be taken from them unless they are captured while fighting with their arms in their hands, by sudden surprise or disabling wounds. When Indians first surrender or come upon a reservation, they anticipate being disarmed, and make their preparations in advance. They caché most of their best weapons, and deliver up only the surplus and unserviceable. The disarming of Indians has in almost every instance on record proved a farcical failure. Let me cite the case of the Cheyennes who surrendered in 1878. They were searched with the greatest care when they were confined, and, it was believed, with fullest success. Yet when they broke out of prison at Fort Robinson, Neb.,[26] they were well armed with guns and knives and ammunition. Doubtless their weapons had been taken apart and the pieces concealed by the women under their clothing, for weeks prior to the outbreak. It is unfair, furthermore, to deprive the Indian of the means of protecting his home and property against the white scoundrels who, armed to the teeth, infest the border, and would consider nothing so worthy of their prowess as the plunder of ponies or other property from unarmed Indians just beginning to plant crops or raise stock. So long as white thieves roam the country, so long should the Indians at San Carlos be allowed to carry arms for their own protection."

To all of which every right-thinking man will say amen. You observe that Crook goes on the assumption that

THE APACHE IS A HUMAN BEING,

after all. That's one of the reasons why Arizona is down on him—that is, the bloviating element of Arizona; it is a significant fact that the old Arizonians, the men who have been there ten, fifteen or even twenty years, believe in Crook. It is the later rabble that hounds him. And that belief in their humanity

---

[26]January 9, 1879.

(not humaneness) is bound to win. We saw the same proposition fought along to the death of slavery. Crook has no silly sentimentalism, no maudlin mercy; but he knows what can and what cannot be done.

Pursuing this subject, Crook says:

"In dealing with this question, I could not lose sight of the fact that the Apache represents generations of warfare and bloodshed. From his earliest infancy he has had to defend himself against enemies as cruel as the beast of the mountain and forest. In his brief moments of peace, he constantly looks for attack or ambuscade, and in his almost constant warfare no act of bloodshed is too cruel or unnatural. It is, therefore, unjust to punish him for violations of a code of war which he has never learned, and which he can with difficulty understand. He has, in almost all his combats with white men, found that his women and children are the first to suffer; that neither age nor sex is spared. In the surprise and attacks on camps women and children are killed in spite of every precaution; nor can this be prevented by any orders or foresight of the commander, any more than the shells fired into a beleaguered city can be prevented from killing innocent citizens or destroying private property. Nor does this surprise the Apache, since it is in accordance with his own custom of fighting; but with this fact before us we can understand why he should be ignorant of the rules of civilized warfare. All that we can reasonably do is to keep him under such supervision that he cannot plan new outbreaks without running the risk of immediate detection; for these *new* acts of rascality, punish him so severely that he will know we mean no nonsense. As rapidly as possible, make a distinction between those who mean to do well, and those who secretly desire to remain as they are. Encourage the former and punish the latter. Let the Apache see that he has something to gain by proper behavior, and something to lose by not falling in with the new order of things. Sweeping vengeance is as much to be deprecated as silly sentimentalism. The Chiricahuas of to-day are no worse than were the rest of the Apaches—6000 in number— who were driven upon the reservation in 1873. The task of managing that number was more formidable than that of looking after the Chiricahuas can ever be; but it was accomplished without any trouble, except such as was stirred up by greedy white men. Many of the Apache chiefs of that day were sullenly opposed to the new order of things. They were ferreted out and broken of their power for mischief, while those who favored the ways of civilization were supported by every influence we could bring to bear."

Well, Crook and his 500 captives got back safely. The campaign had lasted from May 1st to June 9th, and was in every respect a glorious success. The prisoners were put upon the Reservation, and the influences of civilization were brought to bear on them. For two years there was not an outrage or a depredation of any kind committed by an Apache. The civilizing powers had much to contend against, in ways of which I will speak hereafter; and at last there came—incited by a thousand little causes—the outbreak of a year ago.

LUM.

Los Angeles Times, April 22, 1886:

THE MEXICAN FRONTIER

General Miles Urges an Appropriation for Its Better Protection.

[SPECIAL DISPATCH TO THE TIMES.]

FORT BOWIE, A.T., April 21.—General Miles wrote to Washington to-day, recommending that Congress appropriate $200,000 to strengthen the present posts and to establish new ones along the Mexican line, from Fort Bliss, Tex., to Fort Huachuca, A.T. For 250 miles the boundary is unprotected by a single post, and it is between these points that all the Apache raids from Mexico have come up. He adds a suggestion that the Government may be obliged to call on the Mexican Government to remove the hostiles now in the Sierra Madres so far to the interior as no longer to be a menace to the United States.

LUM.

A VAIN PURSUIT.

FT. BOWIE, A.T., April 21.—Captain Dorst returned yesterday, from an unsuccessful pursuit of Geronimo.

[*Probably by Lummis*]

Los Angeles Times, April 23, 1886:

MILES'S ORDERS.

How He Proposes to Hunt Down the Apaches.

[SPECIAL DISPATCH TO THE TIMES.]

FORT BOWIE, A.T., April 22.—Gen. Miles to-day issued general field orders No. 7, distributing the Territory for thorough patrolment. Signal detachments will be kept on the tops of the highest peaks to communicate the

---

[27]The use by Miles of the heliograph for signaling purposes during the Geronimo war is well known. Not so familiar, however, is the fact that the heliograph had been used experimentally and tactically in the southwest since 1882—four years before Miles arrived—or even earlier.—Dan L. Thrapp, "General Miles and the Heliograph," *The Wrangler,* San Diego Corral of the Westerners, Vol. 2, No. 4 (September, 1969). The other provisions of Field Orders No. 7 merely restated traditional practices of troops in the theatre.

movements of the hostiles, and between the camps,[27] infantry will be used in constant hunting through the mountains, occupying the passes, etc. A sufficient number of reliable Indians will be retained for trailers etc. Cavalry will be used in light scouting with sufficient force always ready for instant, vigorous pursuit. To overcome the hostiles' advantage in relays of horses, commanders will dismount half their men and send the lightest and best riders in pursuit till all the animals are worn out. Thus the command should in forty-eight hours catch the hostiles or drive them 150 to 200 miles, if the country is favorable for cavalry, and horses will be trained for the purpose. Commanding officers will thoroughly learn the topography of the section under their charge, and must continue the pursuit till capture or till sure that a fresh command is on the trail. All camps and movements will be concealed as much as possible. To prevent the hostiles getting ammunition, every cartridge will be accounted for, and all empty shells destroyed. Field reports must be made thrice monthly.

LUM.

———————

Los Angeles Times, April 25, 1886:

The Apaches at it Again.

TOMBSTONE, A.T., April 24.—A report from Magdalena, Sonora, announces that the Apaches attacked the Bado Seco ranch, twenty-five miles southeast of Magdalena, killed three men and two women, and destroyed a quantity of property.

It is also reported that Indians, believed to be a portion of Geronimo's band, attacked a ranch near Santa Cruz, Mexico, and killed a number of ranchers.

Los Angeles Times, April 28, 1886:

The Apaches are on another raid through Arizona. This is Gen. Miles's golden opportunity. Let him close his cordon on them.

———————

Los Angeles Times, April 28, 1886:

GERONIMO, AGAIN.
———————
More Apache Outrages on the Border.

ASSOCIATED PRESS DISPATCHES TO THE TIMES.

GUAYMAS, (Mexico), April 27.—A telegram received here to-day announces the appearance of the Apaches under Geronimo near Calabasas, Arizona. Ten persons are reported killed on the ranches near the latter place.

Dispatches to Governor Torres state that over thirty persons were killed on the ranches near Casita. Troops will go forward by rail to-morrow.

CALABASAS, A.T., April 27.—Thirty Indians raided near here yesterday and killed one Mexican and wounded another. A posse of citizens started in pursuit, but when a mile from town were attacked and driven back.

SOME OF THE VICTIMS.

NOGALES, A.T., April 27.—The wife and child of A.L. Peck[28] have been killed by Indians, and his niece taken prisoner. Peck was captured but escaped. Owen Bros., prominent ranchers, are also reported killed.

GEN. MILES AT NOGALES.

NOGALES, A.T., April 27.—The depredations to-day were in the most thickly settled portion of Pima county, and this is the first raid in that section for ten years. It is believed that the hostiles are committing outrages in revenge for the supposed death of the captured portion of the band recently sent to Florida. Gen. Miles arrived last night on receiving intelligence of these raids, and if the hostiles remain in Arizona, it is stated he will take the field.

---

[28]The raid up the Santa Cruz Valley was the last important depredation within the United States by the Geronimo hostiles. Their attack on the Peck ranch was its most celebrated incident. The ranch was surrounded by Indians, ranch hands were killed, Peck was tied up and "compelled to witness indescribable tortures upon his wife until she died." The ordeal left him temporarily violently insane, and the Apaches, in awe of one so afflicted, freed him. The captured child was not mistreated by the Indians but taken into Mexico, where later she was freed by the Lawton command.—Jack C. Lane, ed., *Chasing Geronimo: The Journal of Leonard Wood May-September, 1886* (Albuquerque, University of New Mexico Press, 1970).

In one of the notorious incidents of the Apache wars, Judge H. C. McComas, left; his wife, right; and their son, Charlie, were accosted on the road from Silver City to Lordsburg. Mr. and Mrs. McComas were butchered by Chatto's raiders and their son stolen away, never to be certainly located again. Photograph courtesy the Arizona Historical Society.

Captain Adna Romanza Chaffee, one of Crook's most valued officers, who went into Mexico with him in 1883. Photograph courtesy of David P. Perrine.

First Lieutenant Frank West, who accompanied Crook into Mexico in 1883, later won a Medal of Honor for an Apache fight in Arizona. Photograph courtesy David P. Perrine.

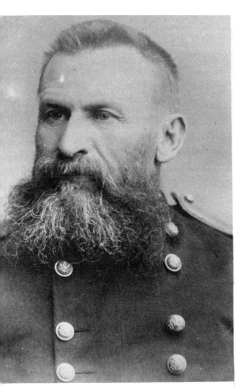

Brigadier General George Crook, about whom Lummis wrote a great deal from Fort Bowie.

Second Lieutenant Charles B. Gatewood, one of the most important and most experience officers in the Apache wars. He was the man who finally persuaded Geronimo to surrender in the summer of 1886. Photograph courtesy the National Archives.

Brigadier General Nelson A. Miles about the time of his Arizona service. Photograph courtesy the National Archives.

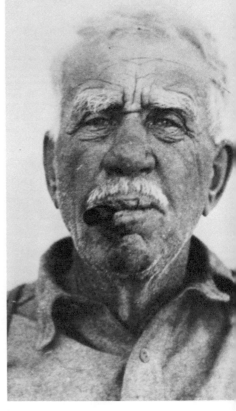

Arthur L. Peck, whose family was destroyed and he taken prisoner by Geronimo's Apaches on their last raid into Arizona. Peck became violently insane at the tragedy and the Indians freed him, in keeping with their customary awe of a crazed man. Peck recovered and in 1927 was living at Nogales, Arizona. This picture was taken about that time. Photograph courtesy the Arizona Historical Society.

CHAPTER IX

Los Angeles Times, May 5, 1886:

CROOK'S POLICY.

The Grey Fox's Creed as to Indians.

HOW THEY ARE AND OUGHT TO BE TREATED

*Common Honesty Necessary—The Question of Using Indian
Scouts—The Chiricahuas the Boss of All.*

SO MUCH HAS BEEN SAID as to Crook's Indian policy, and so ignorantly and maliciously said, that one day when the quiet General had got a little ahead of his work, and was settling down for an hour's rest, I corraled him and asked a definition of his creed as to Indians. It was a subject upon which he was free to talk; and he spoke with an earnestness which showed how deeply his heart was enlisted in this perplexing question. He said:

"The Indian is a peculiar institution. And still, he is a human being. A good many people seem anxious to forget that fact. In dealing with him you must first of all

BE HONEST.

"It is a poor sort of honor, though a popular one, which holds that decency is to be used only toward white folks, and that when you lie to an Indian or swindle him, it doesn't count. No one else in the world is so quick to see and to resent any treachery as is the Indian. You can do nothing toward his management unless you have his confidence. That is true of all men, but particularly true of him. You may fool others as to your intentions, but you can't fool him. He has no books nor newspapers; and so he has to draw on nature for his knowledge. This training has made him wonderfully sharp. He will sit looking at you with the expression of one of these old fashioned crocks. You've seen them—you can't tell by their looks whether there's honey or vinegar inside; and you can tell no more by his face what is behind it. But all this time he is reading you as if you were an open book. He can almost tell from your expression what you had for breadfast! I have known one of these Apaches to go sixty miles out of his way to ask a man the same question he had asked a month before, and see if he would get the same answers. If he doesn't find out whether a man's word is to be relied upon or not, it is a pity.

"Another rule to be followed in dealing with Indians is,

NEVER GIVE HIM A GRIEVANCE

to brood over. Settle the thing up on the spot. These little grievances are trifles in themselves, but they're like little worms boring into a big oak—the last one brings it down. And people at large, who never see anything but the last straw, exclaim, 'Why, what fiends to break out for such a trifle!'

"In warfare with the Indians it has been my policy—and the only effective one—to

USE THEM AGAINST EACH OTHER.

To polish a diamond, there is nothing like its own dust; and it is the same with these fellows. Nothing breaks them up like turning their own people against them. They don't care for the white soldiers, whom they easily surpass in the peculiar style of warfare with they force upon us; but put upon their trail an enemy of their own blood—an enemy as tireless, as foxy, as stealthy and as familiar with the country as they themselves—and it breaks them all up. It is not merely a question of catching them better with Indians, but a broader and more enduring proposition—

THEIR DISINTEGRATION.

By this policy, all their tribal organizations will eventually be broken up. It cannot be done in a moment—it takes time to uproot the institutions of centuries—but it is the inevitable outcome. Now, of course, if you have 10,000 troops and one Indian tied down in the middle of them, you don't need any Indian policy. But when you have to *catch* your Indian, there's where the policy begins to be useful. If we could always put our hands on these fellows, the question of managing them would be simple enough. You can bulldoze a lion when you have him in a stout cage and a red-hot iron in your hand; but when he is on his native heath, then the proposition is different. You can't bulldoze these fellows, either, when they are loose in a wilderness as big as Europe.

"The invention of

BREECHLOADING GUNS

and metallic cartridges has entirely changed the methods and the nature of Indian warfare. It is not many years ago that the Indians were miserably armed, but all that has changed. They are no longer our inferiors in equipment. Instead of bows and lances, they now have the best makes of breech-loading guns and revolvers. In the only kind of warfare to which they can be forced, they are more than the equals of the whites. For white soldiers to whip the Chiricahuas in their own haunts would be impossible. The enormous country which they range is the roughest in America and probably in the world. It is almost utterly bare of anything upon which a white man could exist; but it supplies

EVERYTHING THEY NEED

to prolong life indefinitely. There is no end of the mescal plant everywhere in their territory; and if there were nothing else whatever, the Apache could live very comfortably on the varied products of that wonderful plant. He has no property which he cannot carry along in his swiftest marches; no home to leave at the mercy of his enemies. He roves about like the coyote, as unencumbered and more elusive. He knows every foot of his territory, and can live through fatigue, lack of food and of water, that would kill the hardiest white mountaineer. By the generalship which they have found necessary, they oblige us always to be the pursuers; and unless we can surprise them, the odds are all in their favor.

WHEN IT COMES TO A FIGHT,

we can't see anything of our foe—nothing but the puffs of their rifles. Nothing is exposed but here and there an eye, peering from behind some rocks. *You* can't see that eye, but those fellows, with their marvelous vision, will see your eye at a hundred yards. No white man can take advantage of the ground as they do. Our soldiers have to expose themselves, since they are the attacking party. As a sample of their fighting, look at the time when those Mexican banditti attacked Crawford. When they first fired into the camp, in which our scouts were sleeping, they wounded two men in bed. Then the scouts took to the rocks, and after that, in an hour and a half or two hours of hard firing by 150 Mexicans, not one of those scouts got

SO MUCH AS A SCRATCH.

"When the Mexicans shot down Crawford, a few of the nearest scouts dropped 8 out of 9 Mexicans who did it, at a volley. Then they were checked by Maus, or they'd have wiped that Mexican force off the face of the earth.

"No, to operate against the Apaches we must use Apache methods and Apache soldiers, of course with white soldiers along. The first great difficulty is to discover the whereabouts of the hostiles, and this can be done well only by Indian scouts. Their stronghold once located, the next thing is to reach it secretly. The marches must be made with the utmost stealth, and by night. Fires and noise are absolutely prohibited. The Indian scouts must be kept far enough in front and on the flanks to discover the enemy without being seen themselves; leaving no trail whatever, but slinking along from cover to cover. As soon as they locate the hostile camp, they noiselessly surround it if possible, meantime sending back runners to us. We make forced marches by night, come up and attack the hostiles, if they have not already flown. We kill a few—and the rest escape among the rocks and canyons. It is impossible to pursue them; for every rock may hide an Apache at bay; and with his breechloader he can kill as many pursuers as he pleases, himself secure. Then there is nothing for us to do but to return to our base of supplies, wait until the hostiles begin to feel secure again, and then repeat the same tedious operation. A single element of utmost precaution neglected, and failure is certain."

"AS TO THE APACHE SCOUTS,"

said Gen. Crook, further, "they are indispensable. We could have made no
progress without them. I first began using them in 1872, and have used them
ever since. Nothing has ever been accomplished without their help. 'Have they
acted in good faith?' We have every assurance that they have. They followed
the hostile trails almost as well as bloodhounds. It is nonsense to think white
trailers could have done the work. These white mountaineers howl against the
Apache scouts—but how does it happen that they and the cowboys together
never killed a hostile? The scouts have put in on the chase the most tremendous
work men ever did; and when they succeeded in catching up with the hostiles,
they have invariably fought well. But there is a great uproar because

THE CHIRICAHUAS

have been employed as scouts. There is a reason for their employment. A good
many people think that an Indian scout is an Indian scout, and that that is all
there is about it; but it isn't so. Almost any Indian scout is better in his peculiar
line than a white man; but the ordinary Indians, and even the ordinary
Apaches, are totally unable to cope with the Chiricahuas. The White
Mountain Apaches come nearest to it, being mountain Indians also. But the
Chiricahuas are matchless. How much this may be due to their having been on
the warpath more than the others, and never having been thoroughly thrashed
I cannot say. The warpath is what brings on Indian out—puts the keen edge on
him. At all events, the Chiricahua

IS 'THE BOSS,'

and all the other tribes acknowledge it [This is corroborated by every one who
has had to do with the Apaches. It is a singular fact that for nearly a year the
Apaches on the Reservation have not dared to send their hunting parties out
into the surrounding mountains for fear the renegades might slip up from
Mexico and swoop down upon them.—L.]. We have Tonto, Yuma, Mojave,
San Carlos, White Mountain and Warm Spring [all Apaches (*sic*)] scouts, as
well as Chiricahuas, and all are good; but the Chiricahuas far surpass the
others. They have done mighty effective service all through. When Maj. Davis
made his first expedition, he took a command of cavalry along with his scouts,
but on his second he left the cavalry behind. He found it only an incumbrance.
It was good cavalry, too, but couldn't keep up with the scouts, and never so
much as saw a hostile, though the scouts overtook and killed several."

Capt. Dorst said to me upon his return from the campaign in Mexico,
"It is

SIMPLE IDIOCY

to try to do anything down there without some Indian scouts. No one else knows anything about that untraveled country; no one else can follow a trail as they do, and no one else can stand so much fatigue. My scouts will start at the bottom of a steep mountain, 1500 feet high, and go on a trot plum to the top without stopping. {I have seen them do this.—L.}. There isn't a white man alive who could run 50 yards up the same pitch without stopping to catch his wind. I have been climbing the mountains of Colorado, New Mexico and Arizona for the last seven years, but I can't keep in sight of these fellows when they start. Everyone—the other tribes included—admits the superiority of the Chiricahuas over all other Apaches."

I may add that the howl against Apache scouts comes largely from alleged white scouts of the Frank Leslie stamp, who want the positions for themselves. But the real white scouts—men like Al. Seber[1] or Frank Bennett[2]—cheerfully admit that Apache scouts are indispensable.

LUM.

————

Los Angeles Times, May 6, 1886:
The Killing of Capt. Crawford.

EL PASO, Tex., May 5.—The report of the result of an investigation of the killing of Capt. Crawford has been forwarded to Washington by United States Consul [J. Harvey] Brigham, stationed at Paso del Norte, Mexico. The report will indicate to the State Department that the killing was accidental, but a private conversation with Consul Brigham indicates that he is greatly dissatisfied with the result, and has a firm conviction that the attack on the United States forces was maliciously and knowingly made.

---

[1]Al Sieber (1844–1907) was perhpas the greatest chief of scouts on the Apache frontier. He was not at Fort Bowie while Lummis was there because he had been assigned to the San Carlos Reservation, but he was a figure of moment in most of the important events in Apacheria for twenty years.—Thrapp, *Al Sieber, Chief of Scouts.*

[2]Bennett (c.1852–post 1890) is the subject of an article by Lummis below.—Thrapp, *Dictionary of Frontier Characters.*

CHAPTER X

Los Angeles Times, April 28, 1886:

APACHE ANTICS.

Social Amenities of Bucks and Belles.

DANCING MORE COSTLY THAN ICE CREAM.

*The Army Mule's Virtues as Related by a Driver
Who Swears Not at All.*

[STAFF CORRESPONDENCE OF THE TIMES.]

IN THE FIELD, April 23, 1886.—Besides his toying with the seductive pasteboards, the Apache buck has a national game of his own. It is not so wildly exhilarating as our base ball, nor so shinfully disastrous as Albion's cricket—but it is large enough for him. It is

THE GAME OF NA-JOOSE.

To a man up a mesquite bush, this recreation looks about as rational as shaving the soles of your feet; but Mr. Apache is content to stay with it all day. Upon a smooth bit of ground he builds two tiny straw-piles, 35 or 40 feet apart, and each with a little U-shaped depression on either side. The other machinery of the game consists of two poles, 15 feet long, and looking for all the world like cane fishpoles (each is made, however, of three straight willow sticks, most artistically spliced); and a six or seven-inch hoop, its diameter traversed by a stout cord. The two players stand elbow to elbow, about 20 or 25 feet from the straw-pile toward which they are facing; and drop their poles till the taper tips bend upon the ground, the butts being held six or eight inches higher. One takes the hoop, holds it down between the poles, and with a deft toss sends it rolling forward. As it nears the strawpile, both players pitch their poles forward; and if it chances that the hoop falls across the poles near the butts, you will hear an approbative shout. Upon a closer look, the poles prove to have a series of notches cut along near the butt; while the hoop is similarly notched, and its cross-string has little thongs drawn through its strands at certain intervals. The point of the game lies in which point falls upon which; and there is really room for much skill as well as a sufficiency of luck.

There also came under my observation

A GIRL'S GAME

even less maddening in its allurements. I don't know what they call it, but it appears to have "come without calling." They build a four-foot circle of fist-sized stones, with three entrances. In the center of the circle is a flattish

stone six or eight inches across. The three or four players squat around outside
the circle, and each in her turn grasps three sticks, so shaped as to fit together
into a round piece about eight inches long. She whacks these down smartly
upon the central stone, tallying according to the distance and direction in
which the sticks fly, and marking her score by moving a twig so many stones
ahead in the circle. I don't imagine that nervous prostration is a fashionable
disease in Apachedom.

On the evening before Gen. Crook's 77 Chiricahua captives were shipped to
their Florida prison, they held a

GRAND FAREWELL DANCE—

a very select affair—under the floor-management of Chihuahua himself. The
ballroom was a smooth spot on the eastern bank of the arroyo in which they
were camped. It was brilliantly illuminated by a huge bonfire of cordwood.
The hop began at 9 o'clock, and was largely attended by the bronchos and
scouts, not to mention a generous audience from the post. The Indians were all
in their best toggery, and wrapped from the evening chill by bright-hued
serapes, army blankets, and even patch-work quilts. Every participant had also
donned a brand new complexion. The medicine man, hugging a rude little
drum in the crook of his left elbow, was orchestra and conductor, all in one. He
was surrounded by a dense circle of bucks, four or five deep, who danced in
decorous shuffle to his taps, and sang in perfect unison with him. As nearly as I
could catch the plans and specifications of their song, it went on this wise:

> "Hai, la-i, lai, ennay,
> Nay, ennay,
> Nay, ennay;
> Hai nai, nay, nay ennay
> Hai nay, nay, ennay,"

and so on for 213 verses or thereabouts. The bucks did not make any progress
by their stepping, which was of a perpendicular, tread-mill sort.

THE BELLES OF THE BALL,

from simple six to sweet sixteen—for all the girls above infancy
participated—were stationed a few yards higher up the slope than the men;
while the non-combatant old women sat still further back in the shadows'
verge. Every minute or two a pair of damsels would patter giggling down to
the circle of dancers, slap two swains resoundingly upon the shoulders, and
scurry bashfully back. The two beaux thus favored would leave the circle and
join their partners, each unfolding his blanket from his shoulders and sharing
it with his girl. Then they would form in line—a buck at the extreme right,

facing from the circle; then his partner, dancing toward it; next the other buck, facing from; and last of all, *his* girl, facing toward. Thus arranged, they would "sashay" up and down, now advancing clear up to the circle, and now retreating twenty-five feet or so from it. This would be kept up until the chorister stopped to catch his breath, when the bucks would return to the circle, and the girls run, laughing, back to their companions. I don't know that it does the Apache beau much good to escape the ice-cream tribute, since he is in duty bound to give his dulcinea $5 or $10 every night that she dances.

The ball kept up about all night, "and was the social event of the season." It was a weird sight, down among the bleak hills—the leaping, ruddy fire throwing its fantastic shadows about, while those savage merry-makers flitted here and there with jocund but depressing ululation.

MRS. MANGUS.

Among the most interesting of the captives was a fat, well-preserved, ever-smiling old squaw, wife of the celebrated Chief Mangus, and the mother, I believe of nine children. She was generally called Francesca here (she was one of the guardhouse party) but her real name is Sa go Zhu-ni,[1] which signifies "Pretty Mouth." She is an important personage in more ways than one. They say it was she who made the tizwin which led to the present outbreak. But now her ways are ways of peace—and population. She is the Juno of the tribe,[2] and superintends the additions to the Chiricahua census. When Capt. Bourke and I were down at the guardhouse one day, before the arrival of the surrenderers, she showed us a magical flint arrowhead, made from a peculiar stone, found in a particular spot at a special season, and possessed of great virtues in soothing ante-maternal pains. One of the imprisoned squaws had a bouncing Christmas present, and some of the guards ran for the post surgeon. But old Francesca didn't need any foreign appliances or aid, and when the doctor arrived the demand for him was over. The necessary umbilical surgery had been performed with the sharp edge of a—tomato can! Everyone supposed the poor little

---

[1]There is uncertainty about this individual. Griswold lists neither Francesca nor Sa go Zhu-ni, nor any variant of either. Britton Davis said that Mangus's wife was Huera, an expert tizwin maker who had been captured by the Spanish and spent years among them, becoming fluent in their language and thus a power among the Apaches to whom she returned. Griswold lists Mangus's only wife as Dilth-clay-han, a daughter of Victorio born in 1846, who apparently died at Fort Sill. If Mangus had another wife named Francesca, she could have died before reaching Fort Sill and no recollection of her have been preserved, although that would seem unlikely. Huera is pictured in *Frank Leslie's Illustrated Newspaper,* Vol. LXII (July 31, 1886), 380, and her appearance fits Lummis's description of "Francesca" rather well.

[2]Juno in Roman religion was queen of the gods, the wife of Jupiter and protector of womanhood—kind of an early advocate of "Women's Lib." Lummis suggests that Francesca/ Huera also served as midwife on occasion. This is confirmed elsewhere.

newcomer would directly turn up his toes to the daisies; but he didn't mind a little thing like that; and is to-day as bright and vigorous a sprawler as you are apt to find anywhere.

A NON-PROFANE MULETEER.

Among the noteworthy people I have met here, there is one deserving of special mention—Thos. Moore, nominally Chief Packer of the Department of the Missouri, but in fact Master of Transportation of the whole army of the United States. He has an odd, kindly, homely face, twinkling gray eyes, and a droll turn of conversation. But he is a man of clear, original, and often deep ideas. If there is any creed of universal popular acceptance, it is that a mule—and particularly an army mule—can't be properly engineered without a club and plenty of profanity. But here is a man who has more to do with army mules than anyone else in the country; a man who never swears at anything, and who discharges any employe whom he catches raining blows and profanity upon one of the long-eared train. In speaking of his pets the other day, Mr. Moore said to me: "The mule has never been done justice. It is fashionable to disparage him, and people do so without knowing anything about the subject. God

MADE THE MULE ON PURPOSE.

"The horse has that in his nature which shows that he was designed for something more than a servant to man. God saw that man needed a straight-out servant, so he built the mule. And a true servant he is. Almost as soon as he is able to walk, he begins to be used; and his tireless service ends only when the breath leaves his worn-out carcass. You might almost say that he is useful every day of his life. He is always faithful and always reliable. He understands his work, and does it as few men do theirs. The idea that he must be cursed and clubbed to work is a popular idiocy. He does best with kind treatment. Kick? Yes, he has a bad reputation as a kicker, but that arises mostly from faulty handling. I am in less danger of being kicked when among mules than when among horses. A man accustomed to handling horses would be in less danger among them than among mules. A greenhorn, unused to either, would be in more danger among horses than among mules."[3]

As this is at present the only department in which campaigning is being done—and in which pack transportation is therefore needed—Gen. Sheridan sent Mr. Moore down here to organize the trains, and here he still remains.

---

[3]Moore's ideas of mule handling continued to be Army doctrine as long as pack mules were retained by it—until c. 1955. Moore not only was original in his approach to mule transportation, but also was exactly right in his ideas about the animals and how they should be managed as this editor, himself a onetime pack officer with the Army, can attest.

A ROUGH COUNTRY.

A few days before he left, Gen. Crook was telling me something about the difficulties of campaigning in the Sierra Madres. Among other things he said:

"During the campaign of '83, when I went down after the renegades who were lurking in the Sierra Madres, I remember one night when we camped in the bottom of a deep cañon. Next morning at daybreak we began to climb the sides of the cañon, zig-zagging to the top. When we traveled about half a mile on top of the ridge, and then down into another cañon. *We were all day going six miles!*"

I remarked that one of the most striking proofs of the roughness of the country was the fact—of which I had seen mention in reports of the expeditions—that the sure-footed mules were often killed or crippled by tumbling down the mountains. Gen. Crook replied:

"Well, the pack-mules seldom fall unless they are pushed. The old ones, particularly, know mighty well where they can go and where they can't; and when they come to a dangerous place, they will stop short, and you can't budge them with a club. But the others, coming up behind, will frequently crowd them off, and down they go. I remember once, one of our mules

FELL 2500 FEET.

The trail ran along a narrow shelf at the top of a tremendous precipice, and one old fellow got shoved off by his mates. The doctor was down in the valley below, and saw the whole thing. He says when the mule first came over the edge, it didn't look any bigger than a jackrabbit. When it struck on the rocks at the bottom, it exploded like a cannon, and he couldn't find a piece as big as his hand. The brute had all our cooking outfit on his back, too, and we were left in a bad plight. On one occasion a mule in the pack-train with which I was messing got pushed off the trail on the side of an extremely steep mountain. He was loaded with bacon. He went rolling down and disappeared from our sight in a cloud of dust. There were two packers away down the zig-zag trail, and they declare that that mule bounced clear over them, struck on his back on a big rock below, bounced way up into the air, and finally went souse into a deep pool in the river, whence he swam out and went to grazing on the grass beyond! [Lieut. Faison has told me of a similar incident which he witnessed—L.]. Another mule on the same trip went over the edge and broke his back. He was loaded with ammunition."

I know that a good many mules have been lost in this way in that frightfully rough country, which is to be our stamping ground for the next—*quien sabe* how long. I begin to tremble for my own neck, when I think of all the fatalities in the family.

LUM.

CHAPTER XI

Los Angeles Times, May 2, 1886:

THE COWBOY.

A Good Fellow, but No Indian Fighter.

HOW HE FALLS SHORT OF THE MARK.

*Too Much Hoop-La and Not Enough Experience
with "Injuns"—Some of His Exploits.*

AMONG THE FUNNIEST PARAGRAPHS interjected among the red pages of the Apache campaign there has stood one pre-eminent—the vociferous bazoo of the cowboy, howling to be let at 'em. The exuberant knight of the lariat has had to keep his friends implored to keep him down, lest he avalanche himself upon the hostiles, and expunge the Apache race at one fell swipe. By emphasis of profanity and iteration, he has actually caused some good people to yearn for the formation of cowboy companies, in faith that thus the problem should be solved instanter. Some trusting people have even wired to Washington a proposition to that effect.

Now, there are no particular flies on the cowboy. He is a good fellow in his way—and there are worse ways, too. So far from wishing to run him down, I cherish a lively memory of many manly kindnesses at his hands. He is rough—but the primmest of us would somewhat assimilate with so savage a country as that which he ranges. He has virile virtues not a few, and they are as protuberant as his vices. To damn him by wholesale is unjust—and silly, *tambien.* But

AS AN INDIAN FIGHTER

he is a rank failure. He has neither the experience nor the disposition to trail; and when a fight is on he must gallop madly about at a salutary distance, whoop and swing his hat, and promulgate his six-shooter or Winchester to the mortal peril of the circumambient air, but not of anything else in particular. If you deduce hereby that he is a coward, you were never so badly fooled in your life. Probably no class of men is more absolutely contemptuous of death. He will pluck the grim old Reaper by the beard at any time, and never twitch an eyelid. But he wants to see the deal in any game wherein he is to take a hand. In the bar-room broil, where the friendly glass spills an ugly word, the word is echoed by a blow, and the blow gets answered in the instant flash of twenty revolvers—there the cowboy is at home. He "savvies the burro" there. Untrembling as the rocks, you shall see him face the murderous music of the forty-fours, his own barking back defiance. Sieve him with bullet-holes, and he will yet bring down his man. Knowing his ground, you will find him clear grit to the last flutter of his heart.

But it is "the danger that we know not of" which "makes cowards of us all."

KNOWS NOTHING OF INDIANS.

It isn't his business. Unflinchingly as he will face the danger that he can *see,* when it comes to fighting an invisible, an unknown and a mysterious foe, he "isn't there." I don't blame him—it shows his width between the eyes. To gallop or creep through a waste so bleak, so barren and so desolate that it oppresses the senses; amid a vast silence heavy enough to break the heart; seeing no sign of life, yet ever on the nettles of a consciousness that any innocent bunch of bear-grass, any cactus rosette, any lonely rock, may unwarning spit out its little puff of sudden smoke, with a leaden message to your heart—isn't it enough to make any one a bit loose in the knees? The man whose heart doesn't feel, now and then in this warfare, as though all the stuffing had been kicked from under it—he isn't a hero. He merely needs a little seclusion, with possibly a Stockton corset. [1] No, bravery alone isn't an adequate stiffener. It takes experience in this line, and that the cowboy has not. And, as for discipline—why, discipline him, and he would be no more a cowboy. His absolute independence and individuality are himself. If it will amuse anyone to pick up cowboys and heave them at the meteoric Apache, the target won't complain if the missile doesn't. No one will get hurt—save the cowboys perforated by their brethren—and there'll be as much fun as there isn't gore. But don't ask the government to subsidize the c-b. That would be crowding the mourners.

In illustration of how the bovine-baster fares when his wits are matched against those of the crafty Apache, let me relate a few short chapters of true though unwritten history.

HADN'T LOST ANY INDIANS.

Last October, when Ulzanna and his bloody band were raiding through New Mexico and Arizona, hard-pushed by Crook's efficient captains, it came to pass, one day, that the old men and children of the band, with four or five bucks who could fight, swept down by White's rancho[2] twenty-five miles south of Fort Bowie. Camping in the open plain, half a mile from the house and

---

[1] Straitjacket. The reference is to Stockton, California, where an insane asylum was established.

[2] In this passage Lummis apparently was unclear on several points, and, indeed, the record is confusing, but it appears probable that Ulzanna (Josanie) did not lead this raid, but one a month later.—Thrapp, *The Conquest of Apacheria,* 332–39; Lockwood, *The Apache Indians,* 282–83.

in full view of it, they killed two or three beeves, roasted them, ate all they could hold and packed all they could carry. In the night they flitted silently away, unmolested. Inside the strong blockade of that rancho-house were twenty-five or thirty cowboys, weighed down with Winchesters and six-shooters. Did they open fire on the superannuated foe? Nary one. *They* hadn't lost any Indians!

CHOPPED DOWN THE PALISADES.

Ulzanna and his band holed up in the Chircahua mountains, whence they twice tried to break across the broad San Simon valley to Stein's Peak range on the northeast; but were each time deterred by the sight of troops "laying for" them. At last they made a night dash to the westward, across the Sulphur Springs valley. Their stock was on its final limbs; and while the rest rode on across the valley, three or four bucks went to the Sulphur Springs rancho for fresh mounts. There were several *muchachos de vaca* sleeping in the house that night; and the strong, high stockade which protected the horses was securely locked and barred. The Chiricahuas borrowed a hatchet from the woodpile, chopped down enough palisades, took all the horses and vanished. The cowboys did not fire a shot—indeed, I believe they knew nothing of the affair till morning. To one who has wooed sleep under the raucous Niagara of a cowboy's snore, this seems highly plausible.

PONIES, LARIATS AND ALL.

Capt. Crawford chased the remounted renegades through the Dragoon and Mule mountains. Then they whipped square to the left, and made for the Chiricahua peaks again, with Crawford still hanging on their trail like a wolf. The day after they got back to this range he found the spot where, hard-run and with failing stock, they had stabbed every one of their animals to death and scattered on foot amid the rocks. Then he thought he had a lien on them, sure. Just at this time the cowboys of the San Simon valley were getting fixed for the fall round up, and had gathered in force at a rancho near the mouth of White-tail cañon, with their twenty-five or thirty prime ponies nicely shod and ready for the fun. They were warned that hostiles had been seen on the neighboring peaks that day, and that it would be horse sense to put their ponies in the stout corral at night. But no, they wouldn't have it so. They'd like to see any blankety-blank Indians get away with any of *their* live-stock. So they insisted upon lariating their ponies out on the grass, while themselves snoring conscientiously inside the house. When they awoke next morning, rubbed their confident eyes, every last pony was gone, lariat and all, and Crawford's tremendous pursuit found only the swift trail of these fresh horses, sweeping far down into Sonora, where the savage riders were safe. It's a lucky thing that the cowboys slept inside that night. Had they camped out beside the ponies the raiders might have carried *them* off.

These are fair samples of the fashion in which our e.c., the cow-boy, emerges from the diminutive extremity of the cornucopia, when he has to deal with the lightning movements and matchless cunning of the Apache.

Thus much for the cow-custodian. Commoner than he and infinitely more jawsome, is the

TERRITORIAL HOLY TERROR—

the bad man from Sinville. When he looms above the horizon, the inadequate steer-steerer pales his ineffectual fires. I can give you no better diagram of him than by relating a little episode of actual occurrence.

It was on the eve of Crook's departure for the Sierra Madres in the Apache campaign of '83 that one of those companies of citizens was organized in Tombstone to quench all Apachedom. They were of the class of Arizonians who make the noise—there are lots of good people in this Territory, but they are not mouth-workers. I forget the official name of the cohort, but the appropriate current name among disinterested outsiders was

"THE TOMBSTONE TOUGHS."

They were just more than armed, and were going to wind up the war in a week. Going to the line in hay wagons—it is hard to carry a jug in the saddle—they swooped upon the Mexican custom-house, and explained that they were going down into the Sierra Madres to obliterate the hostiles. To their infinite disgust, the Mexican officials were delighted with the idea. Here *was* a pretty how-de-do, a pretty state of things. What! Were they to be not only allowed but fairly encouraged to beard the lion in his lair? It was simply murderous! The Toughs dawdled around the line until it became hideously plain that no one would stop them—and then they went home.

Their next proposition was to

MARCH TO THE RESERVATION

and butcher the 8000 Apaches peacefully farming there—men, women and children. A horrid proposition, truly, in the eyes of all decent folk, Eastern or Arizonian; but still a very popular one in this Territory. The Tombstone Toughs, superbly mounted, galloped north, and camped displayfully nine miles from Willcox, where Crook was then preparing to take the field in person. That was the reason why the Avengers stopped near Willcox. They also took care that Crook should know of their coming. But, somehow, the quiet old soldier refused to be stampeded. He would not dilaniate his undergarments at the news. Nay, he wouldn't send out so much as a

BOY WITH A PEA-SHOOTER

to quell the Gory Exterminators. Soon weary of waiting to be collared, they invaded Willcox, retailing their plan to make a grease-spot of the reservation.

And still "D——d old Crook" refused to pay any attention to them. No one stuck so much as a straw in their way. Cursing the blood-thirsty heartlessness of the army, the Toughs loped away at last to Fort Thomas. Crook, the sly old fox, had telegraphed the commanding officer there to treat them with perfect courtesy and unconcern; and the order was carried out to the letter. There was now but one desperate alternative left the Toughs, and they spurred away toward the reservation. Official word had been sent to the Indians there that while there was really no danger, it might be well to keep one eye peeled, and they did. The Apache is not one of the fellows that have to wait for a kick. And so it was that every possible approach to their houses was guarded. You might have ridden by those rugged hills a hundred times and never have suspected danger; but it is straight goods that if the Toughs had been 100 companies instead of one, they never would have got through alive. They didn't know this, and they

DIDN'T NEED TO.

It would have been a wild waste of wind to tell them. They rode to within a day's march of the boundary hills, camped until their ginspiration gave out—and then tore home, not having so much as seen the Reservation. But they reported that they had been up and looked into things there, and found all was perfectly satisfactory. The Indians there were honestly working very hard, and it would really be wrong to kill them! Selah! Thus ended the gory career of the Tombstone Toughs. It reminded one of that famous campaign in which

> "The noble Duke of York,
> He had ten thousand men,
> He marched them up a hill one day,
> Then—marched them down again!"

I may add, by the way, that if any Arizona cowboy or citizen avenger has killed an Apache raider, the grave is like that of Moses.

PATRONAGE, NOT PROTECTION.

As you have heard the woeful wail of Arizona for protection against the Apaches, did you ever stop to think how many Arizonians would be left pecuniarily in the hole, were the Indian wars forever ended? Did it ever permeate your consciousness as to what Arizona would do were it not for the more than $2,000,000 annually disbursed within her borders by the War Department? Remove that, and wherewith could she hide her nakedness? What industry has she to fall back on? What crutch of manufacture wherewith to hobble? Cattle she has on the oases of her vast deserts but all Arizona can't be a cowboy—and, anyhow, what chance has anyone but the capitalist to turn beef into bullion? "But there's a vast mining interest." True for you. But will you tell me what Arizona mine is to-day making money, over and above expenses and current rates of interest on cost of plant? Nay, tell me how many are even reaching that notch? Please don't all speak at once!

The lonely ranchero honestly wishes this cruel war were over. So does the outlying farmer. So does the menaced miner. Ditto everyone who apprehends the confiscation of his personal—or of a related—scalp. The Arizonians who would be willing, with A[rtemus] Ward, to sacrifice all their wife's relations to perpetuate their chief source of revenue, are undoubtedly in the minority—but they make the majority noise. Ah, the longer the war can be kept up bloodlessly, with just enough menace to excuse the retention of a strong military force, the better it will suit the big part of Arizona. The words of the wail are for "protection," but the air is for patronage. Last January, for instance, some ten companies of troops were brought here from the coast, on Gov. Zulick's allegation that they were needed to keep the citizens from butchering every Apache on the reservation. There was no more danger of such a break than there is that your old cow will shin up your pepper tree tail first. There are not enough white men in Arizona to storm the White Mountain fastnesses, and what there are would no more try it than they would poke their heads in a bear trap. The decent people of Arizona—the many who are ordinarily not heard from—resented the imputation that they could harbor so barbarous an intention; and resented it so vigorously that Zulick denied his own appeal for troops. But the troops came, and here they are yet—and Arizona is milking just so many more teats.

LUM.

Los Angeles Times, May 4, 1886:

THE HOSTILES.

The Signal Corps' Work.

SAN FRANCISCO, May 3.—Gen. Miles has had assigned to his command a detachment of the Signal Corps of the army, and it is intended to station them at conspicuous points on the mountain tops, etc., in Arizona, and from there to signal by means of flags, lights and flashes to scouting parties the movements of the Indians, when observed.

Los Angeles Times, May 5, 1886:

THE HOSTILES.

A Brush with Apaches in the Mountains.

NOGALES, A.T., May 4.—Lieut. [Powhatan] Clarke of Capt. [Thomas] Lebo's troop of the Tenth Colored Cavalry, arrived here at daylight this morning with dispatches from the front. Lebo had an hour's engagement with the hostiles yesterday afternoon in the Piniyo [Penito] mountains, losing one man killed, one wounded, and three Apaches. Being unable to dislodge the hostiles from their stronghold, Lebo withdrew his troops. Troop L of the Tenth and Troop B of the Fourth Cavalry left here at 1 o'clock this morning for Piniyo to aid Lebo in the second attack which it is proposed to make on the Apaches.

CHAPTER XII

Los Angeles Times, May 6, 1886:

A SAMPLE

Of What Apache Campaigns Really Are.

CAPTAIN WIRT DAVIS'S EXPEDITION.

A 90-days' Pursuit of the Hostiles Last Summer—Rugged Traveling, Intense Heat and Other Pleasures.

TO GIVE AN IDEA of the monotony, wearisomeness and general dimensions of this anti-Apache crusade, I have secured a full and accurate account of Capt. Wirt Davis's first expedition into Mexico after the hostiles, during the present campaign.[1]

The command left Fort Bowie July 7, 1885. It was composed as follows, Capt. Wirt Davis, Fourth Cavalry, commanding; [Second] Lieut. [James B.] Erwin,[2] Fourth cavalry; two companies (102 men) of White Mountain, San Carlos and Chiricahua Apache scouts, commanded by First Lieut. M.W. Day,[3] Ninth cavalry; Second Lieut. Robert D. Walsh,[4] Fourth cavalry, and [Charlie] Roberts and Leslie, Chiefs of Scouts; Assistant-Surgeon H.P.

---

[1]What follows is based on Wirt Davis's official report, a copy of which Lummis apparently obtained from Bourke or the post clerk.

[2]Erwin (1856–1924), born in Georgia, was a West Pointer who joined the 4th Cavalry in 1880. He took part in Apache campaigns in 1885–86. Erwin was superintendent of Yellowstone National Park in 1897–98 and after his Phihippines service had charge of relief work in Oakland, California, following the disastrous 1906 San Francisco earthquake. He took part in the punitive expedition into Mexico, was adjutant general for two years, and became a Brigadier General.—Thrapp, *Dictionary of Frontier Characters.*

[3]Day (1853–1927) was born in Ohio, graduated from West Point, was assigned to the 10th Cavalry in 1877 and to the 9th Cavalry the next year, with which he saw much hard and able service against Victorio. In one action he was threatened with a court martial and awarded a Medal of Honor for rescuing one of his troopers under heavy fire after he had been ordered to fall back. At this writing he was a first leiutenant. Day was repeatedly breveted for heroism under fire during the Indian operations. He was active in Dakota during the Wounded Knee incident. One of his later exploits was chasing pirates with the "bamboo fleet" in the Sulu Sea.—Thrapp, *Dictionary of Frontier Characters.*

[4]From California, Walsh (1860–1928) was graduated from West Point and joined the 4th Cavalry in 1883. He served at Fort Bowie, Arizona, Fort Walla Walla, Washington, and elsewhere in the Northwest. He won a brevet for gallantry in actions against the Apaches September 22, 1885, and June 6, 1886. He retired in 1919 a National Guard Brigadier General.—Heitman; *Who Was Who.*

Birmingham;[5] and two pack-trains. They reached Lang's Ranch[6] (in the very southwestern corner of New Mexico) on the 12th, and were there joined by thirty-eight men of Troop F, Fourth cavalry. They left Lang's the following day, and marching via the Sierra Media [Sierra Enmedio] and Dos Carretas creek (Chihuahua), Bavispe, Baserac and Guachineva [Huachinera] (Sonora), arrived at Huépere creek,[7] 107 miles from Lang's, on the 19th. Here they intended to rest, but at dusk on the 20th, a Mexican mail-carrier brought word from the Prefect of Montezuma [Moctezuma] district that the citizens of Oposura[8] had followed 80 or 100 Apaches—mostly women and children—from near the Sonora river, where they had been depredating, westward toward the Teres mountains, Sonora. Upon receipt of this news, Capt. Davis and his command left camp at 5:20 the following morning, and marched thirty-three miles across the Huépere, Madera and Oputo mountains, and camped at six p.m. six miles north of Oputo.[9] The Presidente of Oputo informed them that the hostiles had that day fired upon citizens fifteen miles north, and that twenty-five or thirty citizens had started in pursuit. These pursuers struck Capt. Davis' camp at sunset, and told him that the hostiles had killed a beef and taken it into the La Hoya mountains.[10] As soon as it became dark, he sent out six scouts to locate the hostile camp; and next morning marched his command down the Bavispe river and camped in a concealed position a mile below Oputo. Two of the six scouts came back to him at eight o'clock on the evening of the 22nd, and said that by watching the squaws out gathering cactus they had located the hostile camp on the highest peak of the LaHoya mountains. At midnight Davis left his pack-mules under guard, and with two days' rations marched across very rough country till daylight, which found him nine miles north of Oputo. Here the horses were tied to trees, and all the men

[5]Henry P. Birmingham (1854–1932), born in New York, was graduated in medicine from the University of Michigan in 1876. He joined the army on February 18, 1881, as an assistant surgeon. He was promoted to Brigadier General of the Medical Corps October 2, 1917.—Heitman; *Who Was Who.*

[6]See Note 19, Chapter 2.

[7]Huépari is eight miles south of Huachinera. The line of march was almost due south from Sierra Enmedio, which is below the border south of Lordsburg, New Mexico. The column continued through Bavispe, Bacerac and Huachinera to Huépari, the latter four places all on the upper Bavispe River.

[8]Since 1828 the community of Oposura has been officially named Moctezuma, although it is still commonly known as Oposura according to Paul M. Roca, *Paths of the Padres Through Sonora* (Tucson, Arizona Pioneers Historical Society, 1967). It is on the Rio Moctezuma, an extension of the Rio Nacozari, 40 miles south of the mining town of Nacozari.

[9]Davis's miles are straight-line or map miles, not the mileage actually traversed by the command, which was farther, of course. His march this day was west-northwest.

[10]The La Hoya Mountains do not show on maps available to this writer; neither does the name La Jolla, which more probably is what they were named. But they would have been north of Oputo some 18 miles, although whether to east or west of the Bavispe River is uncertain.

concealed themselves in the timber. In the afternoon all the scouts were sent out on foot through the arroyos and mesquite thickets, to join the four who had remained to watch the hostile camp. At 7 p.m. Davis started out with his cavalry (all dismounted), leaving the horses tied and concealed eight or nine miles from the mountain. Before daylight the command had completely surrounded the hostile camp with the utmost precaution. At daybreak they found, to their chagrin, that the enemy had noiselessly decamped. Fragments of dresses found in the deserted camp were recognized by the scouts as belonging to women of Geronimo's band.

Ascertaining the direction taken by the hostiles, and sending a scout back for the pack-train, Davis quietly marched his force down to the Bavispe river, and camped, hidden in the timber and canebrakes, near the mouth of the San Juan,[11] where the pack-trains rejoined him. The scouts informed him that there was a big spring on the eastern slope of the La Hoyas, where the renegades often stopped six or seven days to roast the mescal, which was abundant there. He sent eighty-six of his best scouts, with five days' short rations, into the La Hoyas to surprise the hostiles, or to follow their trail towards Huépere creek, where he would join them August 1st. With the rest of his command he marched down to Bavispe, and across the mountains to Huépere creek, ten miles southeast of Guachineva [Huachinera], arriving there August 1st, and finding his scouts with good news. On the 28th of July, Bi-er-ley, First Sergeant of Walsh's scouts, had gone into the La Hoyas with a few Coyotero Apache scouts, and had ambushed four hostiles, killing two and capturing four horses, three saddles, bridles and blankets.

The trail of the hostiles now passed three or four miles west of Huépere, bearing south toward the Sierra Madres. On the 2nd of August, Davis sent Lieut. Day and Chief of Scouts Roberts, with seventy-eight picked scouts and all the rations that could be spared, out on the trail; telling Day that as soon as the pack-train returned from Lang's he would proceed with the balance of the command to Nacori.[12] Davis kept in camp near Huépere with Lieut. Walsh, Leslie and twenty-four scouts; and August 3rd, sent Lieut. Erwin with a detachment of cavalry and the two pack-trains to Lang's for forty days' rations. Pending the return of the pack-train, he sent Walsh and Leslie out several times to look for trails, thinking that Chihuahua's band—which had been attacked by Captain Crawford's scouts, June 23rd—might move toward the Sierra Madres in the same general direction which Geronimo had taken. Gen. D.M. Guerra[13] with 500 Mexican cavalry (regulars) had arrived at Bavispe about August 1, either marching against the Yaquis or to protect Sonora from

---

[11]The Rio San Juan joins the Bavispe six miles north of Oputo.

[12]Nácori Chico is 35 miles south of Huachinera. There is another Nácori Chico (and a Nácori Grande) westerly toward Hermosillo, but neither was the place meant by Davis-Lummis.

the hostiles. He marched within five or six miles of Davis, but they did not meet.

On the afternoon of Aug. 7, Lieut. Day and his scouts surprised Geronimo's camp about 30 miles north-north-east of Nacori, killed three bucks, one squaw, one boy and a child, besides capturing fifteen squaws and children, thirteen horses and mules, the blankets, saddles and whole camp outfit. (And yet some remarkably wise journals allege that the Apache scouts won't fight with the hostiles!)

Aug. 14th, the pack-trains returned loaded from Lang's; and on the 15th, Davis and his command started for Nacori, taking part of one pack-train loaded with rations. Arriving at Bacadehuachi[14] (31 miles) the same day, they found [First] Lieut. [Guy E.] Huse[15] with Co. C, 4th Cav., and some Chiricahua prisoners from Capt. Crawford. Next day the prisoners were all sent to Fort Bowie, with Lieut. Huse; and Davis sent for Lieut. Day and his scouts—who were camped between Nacori and Bacadehuachi—and directing Lieut. Erwin to join him with the balance of the command and the pack-train which had been left at Banirancho. Aug. 20, the whole command camped at Ojo Caliente,[16] six miles northwest of Bacadehuachi, and rested for three days. Capt. Crawford had gone into the Sierra Madres in pursuit of Geronimo's scattered band.[17]

Learning from the Prefect of Oposura that twenty or twenty-five hostiles were in the mountains between Oposura and Tepache,[18] Davis started at once, August 23, in pursuit. Crossing the Huasivas [Huásabas] and Oposura mountains, and passing through the towns of Granadas and [Huásabas],[19] the command, on the 26th, arrived at Toni-Babi,[20] in the Oposura mountains nine miles east of Oposura. A vaquero sent out by the Presidente of Granadas

[13]No General D.M. Guerra is listed in *Diccionario Porrua de Historia, Biografía y Geografía de Mexico,* 2nd edition, (Mexico, D.F., Editorial Porrua, S.A., 1965), although that compendium is not all-inclusive. There is a Brigadier General Juan E. Guerra (1838–1919), who was active in the north at this time; he had fought against the French and later would side with Porfirio Diaz before retiring to Mexico City.

[14]Bacadehuachi is 23 miles northwest of Nácori Chico, 33 miles southwest of Huachinera.

[15]Huse (1855–1893) was a West Pointer who had been born in New York State. He joined the 4th Cavalry in 1879, by now was a first lieutenant, resigned in September of this year (1886), probably for health reasons, became a civil engineer in Latin America, and died in Guatemala.—Heitman; Cullum.

[16]Probably the place called today La Canoa, on the Bavispe River.

[17]The main Sierra Madre was to the eastward, along the Sonora-Chihuahua border.

[18]Tepache was 20 miles south-southwest of Oposura (Moctezuma).

[19]Granados (Granadas) is five miles north of La Canoa, on the river; Huásabas is four miles north of Granados. Davis was looping around to the north, presumably to avoid alerting the hostiles. The Oposuras and Huásabas mountains are westerly, between the communities of Huásabas and Moctezuma, or Oposura.

[20]The place today is called Atamillo. It is situated on a fork of the Arroyo Tonibabi.

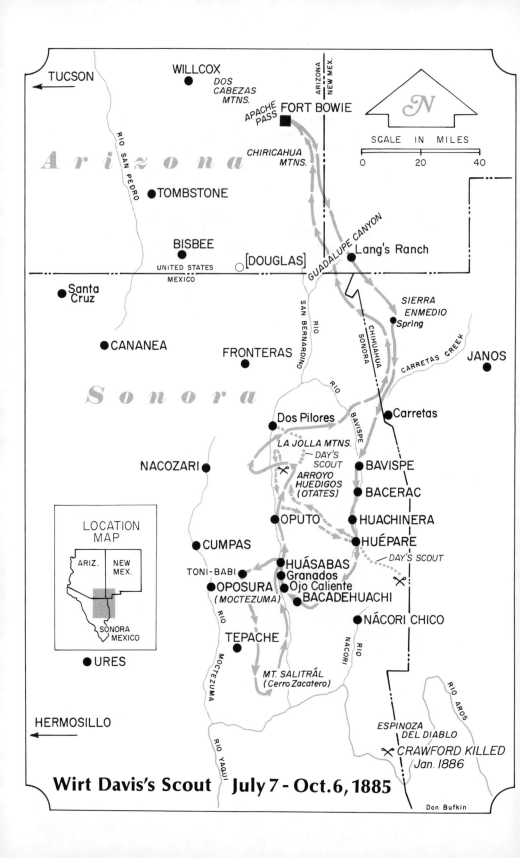

Wirt Davis's Scout July 7 – Oct. 6, 1885

TUCSON

WILLCOX

DOS
CABEZAS
MTNS.

APACHE
PASS

FORT BOWIE

Arizona

RIO SAN PEDRO

CHIRICAHUA
MTNS.

TOMBSTONE

SCALE IN MILES

0   20   40

N

BISBEE

[DOUGLAS]

UNITED STATES

MEXICO

GUADALUPE CANYON

Lang's Ranch

Santa
Cruz

SAN BERNARDINO

CHIHUAHUA SONORA

SIERRA
ENMEDIO
Spring

CARRETAS CREEK

CANANEA

FRONTERAS

Sonora

RIO

JANOS

RIO BAVISPE

Carretas

NACOZARI

Dos Pilores

LA JOLLA MTNS.

DAY'S
SCOUT

ARROYO
HUEDIGOS
(OTATES)

BAVISPE

BACERAC

HUACHINERA

OPUTO

CUMPAS

HUÉPARE

DAY'S SCOUT

LOCATION
MAP

ARIZ.

NEW
MEX.

SONORA
MEXICO

TONI-BABI

OPOSURA
(MOCTEZUMA)

HUÁSABAS

Granados

Ojo Caliente

BACADEHUACHI

URES

RIO MOCTEZUMA

TEPACHE

RIO NACORI

NÁCORI CHICO

RIO

MT. SALITRÁL
(Cerro Zacatero)

HERMOSILLO

RIO YAQUI

RIO AROS

ESPINOZA
DEL DIABLO

CRAWFORD KILLED
Jan. 1886

Don Bufkin

showed them a trail made by twelve or fifteen hostiles the day before. After following this trail a short distance with a few scouts, Davis found that the hostiles were going slowly. He decided to leave his pack-train at Toni-babi, take a few mules, travel all night, and lie concealed by day, until he should surprise the hostiles, who were traveling toward the Espinosa del Diablo,[21] the roughest region in all Mexico. That night Captain Davis received a note from General Guerra, asking a meeting in the morning. Next morning Guerra came over with his officers and an escort, all neatly uniformed. He said his troops had had a fight, July 25, with the hostiles Davis was after, in which one hostile and two Mexicans were killed. He wanted some of Davis's scouts to go with him as trailers, but they declined (I'd like some hydrophobic journals to stick a pin right here, at this Mexican testimony to the value of the Apache scouts). He said he would order his captains to report to Davis for any service desired, and would render every assistance in his power.

Davis's command immediately took up the hostile trail and went to the scene of the fight between the Mexicans and the renegades; and thence to the top of Mt. Sa[li]tral,[22] twenty miles southeast of Tepache, and about opposite the junction of the Haros and Bavispe rivers. Here the hostiles, seeing that they were closely pursued, had taken their knives and killed their stock— sixteen horses and mules, besides eleven which they had similarly killed at scattering points along the same trail. Here the scouts got three saddles, five blankets, one horse and one mule (the latter slightly wounded.) It was, however, impossible to pick up the trail beyond this point, as the hostiles, after killing their stock, had scattered on foot, and a heavy rain had since fallen, entirely obliterating their footprints. Davis camped at Palmiramenes, east of Mt. Salitral, and stayed there the next day, Aug. 31. On the latter day he sent a scouting party around Mt. Salitral, but they could strike no trace of the scattered hostiles. From the beds, etc., which were found on top of the mountain the scouts learned that the band they were then pursuing was Chihuahua's, consisting of twelve bucks, one squaw and one or two boys.

Thinking the band would finally leave Sonora and raid into Chihuahua, via the Teres mountains, Davis decided to march between these mountains and Nacosari.[23] He started September 1st, moved slowly and cautiously northward to the Bavispe river, and marched up that stream; through Granados and

[21]The Espinosa del Diablo, or "spine of the devil," is within the southerly loop of the Aros (Haros) River, directly on the Chihuahua-Sonora line, northerly from La Juna, Chihuahua. It is approximately 80 miles southeast of Oposura.

[22]Today's Cerro Zacatero.

[23]It seems more probable, in view of the geography, that Davis gave up on Chihuahua's scattered band and now determined to head back north and try to intercept another group, perhaps even Geronimo's, reported operating in that region. Nacozari (Nacosari) was about 40 miles north of Oposura (Moctezuma), and the Teres Mountains probably were today's Pilares de Teras, seven miles northeast of the place today called Pilares de Teras, on the river, 27 miles northeast of Nacozari.

Huasivas, camping, September 6th, in an arroyo ten miles north of Huasivas. Next day Lieutenant Cole[24] arrived from Huépari, with rations and commissary funds, and was sent back with the pack-train, on the 9th, to Lang's Ranch. On the 12th Davis started out again, by way of Oputo, and camped next day in the Cañon de los Huedigos,[25] sixteen miles north northeast of Oputo. The cañon was deep, full of large trees (huedigos) and dense undergrowth, and afforded a fine place of concealment for the troops. There was water and grass in the side canyons, so that the horses could graze unseen. Davis had written General Guerra to have the Mexican cavalry at Cumpas[26] and Nacosari keep a sharp lookout, as the dismounted hostiles were apt to sneak over there to steal horses and mules. On the 14th Davis sent Lieutenant Day and his company of scouts northward toward Dos Pilores,[27] on the Bavispe river, to see if he could find any trails leading east; also dispatched parties of Lieutenant Walsh's scouts westward on a similar mission. On the 18th Lieutenant Day returned, having found no trails. The same day, Davis received a note from P.G. Hatcher of the American Ranch, between Nacosari and Cumpas, stating that on the night of the 14th the hostiles had stolen fourteen horses and mules from his ranch. On the 19th Davis started toward Nacosari. After marching four or five miles, he found a man and woman concealed in a ravine. They were Americans from Tombstone, on their way from Nacosari to some mines in the Nacosari mountains. On the preceding afternoon their party—four men and the woman—had been attacked by hostiles. One American was killed, and all their burros and property taken. Taking the man and the woman, Davis camped immediately in the mountains at Nogalitos Spring, in concealment. In the afternoon he left Lieut. Erwin there, with the cavalry, eighteen scouts and a pack-train; and taking eighty-four scouts and a pack-train with twenty days' rations, pushed forward on the trail. They found and buried the dead American, but could discover no trace of the two who had run away. Near the dead man were four empty shotgun cartridges. The brave woman, when the man was shot down, had snatched the shotgun, and put four loads of buckshot at the hostiles, while her two valiant countrymen were running for dear life, leaving her to shift for herself. But she was luckily equal to the occasion. Under fire from the hostiles, she went to the dead man's body, snatched his belt, full of cartridges, and his Winchester, and retreated in good order. When Capt. Davis found her and the one American man who had stood by her, she had the cartridge-belt around her waist, the double-barreled shotgun in one hand, and the Winchester in the other. Davis thinks one of the hostiles carries some

[24]This might have been James A. Cole (1861–1932), 6th Cavalry, a New Yorker who entered West Point from Wisconsin and joined the 6th in 1884.—Heitman.

[25]Today's Arroyo Otates.

[26]Cumpas was on the Rio Nacozari, between Oposura and Nacozari.

[27]About 30 miles north of Oputo.

buckshot in his body yet, from this heroine's gun. If the blowhard fraternity of Arizona could be vaccinated from Mrs. Belle Davis, we might hear of their killing a hostile at least once or twice a decade.

Pursuing the trail, which was evidently made by twenty or twenty-five hostiles, and following it day and night, the command arrived, at noon on the 22nd, in a rugged cañon on the summit of the Teres mountains. The hostiles were traveling rapidly, and leaving a guard behind to watch for pursuers. Davis sent out Sergeant Cooley (an Apache of Day's command)[28] with nineteen picked scouts and several field-glasses, to keep two or three men ahead, travel slowly and cautiously—keeping off the trail—and locate the hostiles, who were evidently not far ahead. Davis himself was to follow in two or three hours, with the rest of the command. He started at 3 p.m., directing packer Patrick to follow with the pack-train at 4, but to stop if he heard firing. Davis had marched a mile and a half, when he heard four or five rifle shots, three or four miles ahead, and apparently on the trail. He hurried his scouts ahead to the firing; and just before sunset they overtook and speedily routed the hostiles, who fled into the mountains. It seems that Cooley had struck the rear-guard of the hostiles and captured their horses; after which he imprudently pushed two sergeants forward on their trail. On a ridge covered with dense chapparal, about a mile from where they had captured the horses, the two sergeants literally walked into an ambush which the fleeing hostiles had made. One sergeant, Cooley's brother, was killed, having been *within ten feet* of the man who shot him. That will give you an idea of the skill of an Apache ambush. The other sergeant escaped. In the ensuing fight, Davis's scouts showed great spirit, stripping for the fray according to their custom. One of them was slightly wounded in the thigh. One hostile was killed, and one or two others must have been wounded, as bloody bandages and considerable blood were found on the trail next day.

The command camped after dark on the scene of the battle. Next morning, September 23, burying the dead sergeant, they resumed the pursuit, going through the Teres mountains, across the Bavispe river, and northeastward towards Chihuahua. On the 25th Davis sent a courier to Lieut. Erwin at Nogalitos, to go with his men to Dos Carretas creek, Chihuahua;[29] also a message through the Presidente of Bavispe to Capt. Crawford, recounting the fight in the Teres mountains, the direction the hostiles had taken, and Davis's belief that they had all left Sonora and gone over into Chihuahua. The trail, which Davis followed steadily over steep mountains and across savage cañons,

---

[28]This man and his brother may have had some connection with Corydon E. Cooley, famous white scout and Apache-man of Arizona, who had married into the White Mountain Apaches.—Wharfield, *Cooley*.

[29]Northeast of Bavispe; the road to Janos follows it for a space. It is directly south of Enmedio Mountain.

debouched, September 25th, on the Janos plains, six or seven miles north of Dos Carretas creek. The hostiles, though now well mounted, were not over four hours ahead of their pursuers. As soon as they reached the plain, they set out on a gallop, and, having gone eight or ten miles thus, nearly due east, scattered in all directions. At 8 o'clock that night, Davis camped on Dos Carretas creek. Just before dawn on the 26th, Lieut. [Charles P.] Elliott[30] arrived with fifty scouts from Capt. Crawford, who was camped on the creek, four or five miles above Davis, and had received the latter's note by midnight. As the command started out towards the Sierra Media [Enmedio] to cut the hostile trail, Crawford joined them. After much time and labor in following the trails of single horses, they struck the main trail of the hostiles in the Raton mountains,[31] ten miles north of where the fugitives had scattered. From here, Davis sent a courier to Lang's with a dispatch advising General Crook that the fugitives were headed toward Guadalupe Cañon.[32] The trail now led out four or five miles into the plain, and thence back again into the Ratons. On the 27th, Lieutenant Elliott got back from the Sierra Media, and rejoined Davis and Crawford; and the whole command moving as rapidly as possible, reached Guadalupe Cañon the next day. The hostiles had crossed the cañon in the morning, about five miles above the cavalry command stationed there. Before daylight, next morning, Captain [John W.] Martin[33] (Fourth Cavalry), with Troop H and some scouts, started in pursuit.

Davis's scouts had now followed the hostile trail rapidly for 270 miles. Twenty or thirty of them were nearly exhausted, and many more had torn moccasins and lacertated feet. Captain Crawford and his scouts were comparatively fresh, having been in camp on Dos Carretas creek for two or three weeks; so Davis sent them on the trail. He also dispatched a courier to General Crook, stating the situation and his opinion that the twenty or twenty-five hostiles were headed for the reservation after recruits. That was the belief of Chatto and several other principal scouts.

Between the Nacori mountains and Guadalupe Cañon these hostiles had killed, when hard-pressed, no less than 30 of their own horses, mules and

---

[30]Elliott (1860–1943) was born in South Carolina and graduated from West Point, joining the 4th Cavalry in 1883. His service in the Geronimo war was varied and good and afterward he futilely chased the Apache Kid on occasion. In 1893–94 he performed rescue and exploratory service in the Bitterroot Mountains of the Montana-Idaho country and retired for disability in 1898. He managed marine transportation for an Alaska exploring party, supervised construction on the Navaho Reservation in Arizona, and served with the Quartermaster Corps in the transportation service during World War I.—Thrapp, *Dictionary of Frontier Characters.*

[31]South of Enmedio Mountains.

[32]On the Arizona-New Mexico line, directly north of the Border.

[33]Martin, from New York, attended West Point from 1869 to 1871, but did not graduate, although he was appointed a second lieutenant in the 4th Cavalry in the summer of 1872. He retired a captain February 25, 1891.—Heitman.

burros. Davis felt confident that they had all left Sonora, and were somewhere in Chihuahua or the United States.[34] It has since been discovered that they were near Casas Grandes, Chihuahua.

October 2, Davis sent a courier to Lieut. Erwin, at Dos Carretas creek, directing him to march with the balance of the command to Lang's Ranch, N.M., and await further orders. Oct. 3, as authorized by Gen. Crook, Davis marched the command to Fort Bowie, arriving Oct. 6. There the two companies of scouts were mustered out and paid, proceeding to San Carlos and Fort Apache.

From the time Davis's command left Fort Bowie until it got back to that Post, 90 days, it had marched between 900 and 1000 miles, not counting the great distances traveled by detachments and scouting parties. "But miles," very sensibly says Capt. Davis, "do not convey a correct idea of the work done. The country is high and rugged beyond description. The heat, particularly about Oposura, Tepache and the Bavispe River, was intense and stifling. To cap the climax of our discomfort, heavy rainstorms in August and September drenched the command and everything it had. Mosquitoes and other insects tortured us at night; and by day we were fully occupied in trying to find and circumvent the wily foe. In such heat the diet of bacon, beans and hastily-cooked bread lost its charm. But the hardships were cheerfully borne by all. The scouts behaved zealously and did effective service through the whole campaign."

He also acknowledges his indebtedness to the Presidentes of Bacadehuachi, Guachineva, Oputo and Granadas for correct information regarding the movements of the hostiles, and for interest in the success of our troops. They were treated well by Gen. Guerra, and by the citizens of Sonora in general.

This detailed story of one expedition among the many of this campaign, gives an idea of just how easy it is to run down the wild raiders. I would like to see about 200 newspaper men whom I could mention forced to join such an expedition. Their brains might then get rid of a little of their present military flatulence.

LUM.

Los Angeles Times, May 7, 1886:
Senator Fair's Plan to End Apache Troubles.

ASSOCIATED PRESS DISPATCH TO THE TIMES.

WASHINGTON, May 6.—Senator [James G.] Fair, of Nevada, has written the

---

[34]That is, the first band he had chased probably were in Chihuahua, the second in the United States. The scouts under Britton Davis and Al Sieber tracked one fragment of the band of hostiles to near Casas Grandes, Chihuahua, thence northerly toward the New Mexico line.—Davis, *The Truth About Geronimo.*

following letter to the Secretary of the Interior, setting forth a plan for the settlement of the Apache question:

UNITED STATES SENATE,)
WASHINGTON, APRIL 10)

To Hon. L.Q.C. Lamar, Secretary of the Interior—DEAR SIR: I beg to inclose the pamphlet of which I spoke to you yesterday. By examining the map you will see the exact position of Santa Catalina Island. It is isolated and too far from the shore for any ordinary boat to reach it. My idea is that if the Apaches were put on this island they would require no guards, and all that would be needed would be a small steam-tender and crew to run between the island and the port of Wilmington [California] for the purpose of communication and supplies. This would, in my opinion, set the Apache question at rest forever, would save the country many valuable lives and the government millions of money. The title to the island is perfect. As near as I can remember, the price asked by the owners is either $3 or $4 an acre. Owing to the location of the island, the climate is as mild and salubrious as can be found anywhere in the world. Should you look with favor upon my suggestion as to the value of this island for the purpose named, I shall be glad at any time to go more into detail with you upon the subject. I am yours,

Very truly,
Jas. G. FAIR.[35]

Senator Fair, when spoken to upon the subject, said that he really believed the Indians would be pleased with Santa Catalina after they got there, although they might possibly fight against going. The island is situated in the Pacific ocean, about fifty-five miles southwest of Los Angeles. The climate, he said, is delightful, and the fishing excellent. Should the island be purchased and the Indians placed thereon, it will forever put an end to annual summer raids of these Indians upon unprotected white settlers and their families, and the government should take some steps in this direction looking to the protection of its citizens.

The *Los Angeles Times* commented: . . . "The chief objection to the scheme is that it would interfere somewhat with the usual summering expeditions to the island. A good many Los Angeles people have a fancy for going over there in the dog days to camp, hunt, fish and bathe. With Geronimo installed. . ., the popularity of the resort would diminish amazingly. . . ."

[35]Fair (1831–1894) was born in Ireland and reached California in 1849, engaging in mining until 1860, when he went to Virginia City, Nevada, where his mining and other endeavors proved lucrative. He served in the U.S. Senate from 1881–87 and died in San Francisco.— *BDAC.*

Tom Moore, whom Crook considered the greatest mule packer in the world, and therefore equal to half of his command on a field scout. Notice Moore has both reins on the mule's near side, indicating he had taught it the difficult (for a mule) skill of neck-reining. F. Jay Haynes photograph, Haynes Foundation.

The Arizona cowboy has all the courage in the world when engaged in his familiar pursuits, but these do not include trailing and fighting Indians, according to Charles F. Lummis. Frederic Remington pictures him above and on the following two pages.

Arizona cowboy.

Arizona cowboys.

(*Facing page*) Moctezuma (Oposura), Sonora, taken about 1900, looking no doubt about as it did in 1886 when it was visited by Wirt Davis and many another American soldier hunting Apaches. Photograph courtesy the Arizona Historical Foundation.

Typical Apache rifle position, made more secure by piling rocks up to form a rude breastwork. This one is on Loco Hill, Sonora, where Loco's band of fleeing Chiricahuas held off Captain Rafferty's troopers. Sierra Enmedio in the background. Photograph by the author.

Wirt Davis, who was in many Indian fights, including the Geronimo campaigns, became Colonel of the 3rd Cavalry. Photograph from *A History of the Third United States Cavalry,* by Ralph Conrad Deibert (Harrisburg, Pennsylvania, Telegraph Press, 1933).

Frank Bennett, scout, whom Lummis considered one of the best in the Southwest. His experience extended from the Plains to well into Sonora. Photograph Arizona Historical Society.

CHAPTER XIII

Los Angeles Times, May 7, 1886:

A REAL SCOUT.

Frank Bennett, the New Mexico Frontiersman.

BRIEF SKETCH OF AN EVENTFUL LIFE.

*Sixteen Years on the Frontier, as Chief of Scouts
Under Hatch, Mackenzie, Crook and Others.*

THE TIMES has already noted the presence here of Frank Bennett, the noted New Mexican scout. He has had an eventful life, though still a young man, but is reticent and modest in speaking of himself—a violent contrast to the class of border toughs whose highest dream is to pose as scouts, and who no sooner reach a town than they promulgate their mounts to the aural paralysis of the population. Mr. Bennett is about five feet, seven inches in height and weighs in the vicinity of 160 pounds—a trim, powerful, alert frame, which no observant quarreler would be in haste to tackle. His face is intelligent, keen and bronzed with healthful exposure. A heavy mustache of blonde tendency adds drooping graces. His hair is thickly sprinkled with gray, but not from age. He has not passed his 35th year.

A sketch of the salient points in his career, as they have come into the possession of a TIMES representative, will not be found uninteresting.

Mr. Bennett was born in Minnesota, where his father was for eight years Register of the United States Land Office. Thence, upon the election of Lincoln, the family moved to St. Louis. In 1870 young Bennett struck out for the wild West, and, the same year, went to the Indian Territory as a scout under Lieut.-Col. [William H.] Lewis, who was killed in 1876 by the Sioux.[1] Bennett served under him throughout the Cheyenne war there. He was present at the fight which the gallant Rucker had with the famous Black Kettle and his band.[2] Out here his campaigning came near being summarily ended. Nearly opposite the little town of Grenada,[3] one day, while the troops were in

---

[1] Colonel Lewis was wounded mortally in action with the Cheyenne Indians (not the Sioux) on September 27, 1878 (not 1876) on Punished Woman's Fork of the Smoky Hill River, Kansas. This was during the Dull Knife campaign, when the Cheyennes drove northward from their hated Fort Reno, Oklahoma exile toward their old country among the Sioux.

[2] What action is meant here is unknown, although the Rucker cited is probably Louis H. Rucker (1842–1906), who enlisted in the Chicago Dragoons in 1861 and rose to become Brigadier General at his retirement in 1903.

[3] Granada, Colorado, is south of the Arkansas River, 17 miles east of Lamar.

camp, the cavalry horses became stampeded and broke, pell-mell, through the tents. The wives of Lieut. [Major Charles E.] Compton and Capt. Chaffee were in one tent, which was knocked down, and Bennett sprang to it just as a band of terrified horses were about to trample it under foot. Seizing a pole he split the onset, the animals passing on either side of the tent without injuring the ladies. But a heavy pin, flying at the end of a lariat, had struck Bennett on the head, and he was stretched senseless upon the ground. He was taken to Fort Lyons [Colorado], and lay there, for nine weeks, on the verge of death. For six weeks he could not even turn over in bed. As soon as his condition would permit he was taken home, and remained there a year. An ugly hole above his left ear remains as a reminder of a chivalrous deed.

Having become fully himself again, Bennett once more turned his face westward, and made the trip via La Jun[t]a to Camp Verde, in the northern part of Arizona, on horseback. After tarrying awhile at Verde, he went down to Prescott, and was there when the Chiricahuas first made themselves felt. Prescott was then about the jumping-off place. Everything had to be freighted in from the coast or the East. Beans were 16 cents a pound, and bacon 50 cents, while flour was firm—well it might be—at $28 to $30 a hundredweight. In 1876, when the Chiricahuas were moved from the Bowie reservation to that at San Carlos, Bennett went to San Carlos and was in charge of the police department until old Victorio broke out in New Mexico in '78.[4] Tired of the comparative inactivity of life on the reservation, Bennett went to New Mexico as Chief of Scouts under Lieut.-Col. [Edward] Hatch of the 9th Cav.[5] Col. Hatch was under every possible disadvantage. The outbreak came like thunder from a clear sky, and he had to take the field instanter, without pack-train and almost without scouts. When Victorio went out from Ojo Caliente, he took

---

[4]Bennett does not seem to have been quite so prominent as Lummis thought; at least his name almost never appears in primary source material on Apacheria at this period, although that in itself is not conclusive. Clay Beauford (Welford C. Bridwell), appears to have been chief of San Carlos police during virtually the entire period Clum was there, resigning September 1, 1877, although Dan Ming, too, occasionally served in some such capacity. It is possible Bennett was hired afterward as chief of police; the record is unclear and has never been adequately researched.

[5]Hatch at this time was a colonel, not a lieutenant colonel. Born in Maine, he was commissioned captain of the 2nd Iowa Cavalry in 1861 and rose to brevet Major General of Volunteers during the Civil War. Afterward he became colonel of the 9th Cavalry and took part in various frontier activities. As commander of the District of New Mexico he was in charge during the Victorio War, where his persistence was marked, if success against the wily Mimbres chief was elusive. Hatch died April 11, 1889, at Fort Robinson, Nebraska, from the results of an accident.—Thrapp, *Victorio and the Mimbres Apaches.*

[6]Hooker, born in New York, was commissioned lieutenant colonel of the 6th California Infantry in 1863 and rose to brevet Brigadier General by 1865. He died in 1883. Lummis's summary of the Victorio activities in New Mexico is fairly correct.—Thrapp, *Victorio and the Mimbres Apaches.*

Capt. [Ambrose] Hooker's[6] whole cavalry herd, killing three of the five colored guards, and wounding another. He had 190 men—Mescaleros, Warm Springs, Chiricahuas and a few Navajoes. The citizens of Hillsboro [N.M.] followed him down to where Lake Valley [then called McEver's Ranch] now is, and had a fight, in which seventeen citizens were killed. Thence Victorio went into old Mexico, via Lake Palomas [Chihuahua]. In '79, Victorio tried to get back to San Carlos to get his wife and old Loco. He was met at Ash Creek by Capt. [Adam] Kramer, who whipped him and turned him back to New Mexico,[7] whence he returned to Mexico. Then 21 companies under Col. [George P.] Buell followed him across the line.[8] Victorio, with the old people, women and children, was in camp, while old Chief Nanay was out with the bucks watching for the pursuing Americans. The Mexican troops, under Gen. Terraces,[9] surprised the camp, killed Victorio and sixty-four others, and carried off the rest to the City of Mexico, where they were put to work in the mines. This was in November [October], 1880. There were no more raids until January, 1881, when **Nanay** came up and shed considerable blood.[10] Then all was quiet till July of the same year, when Bennett was with [Second] Lieut. [John F.] Guilfoyle,[11] who had a roving commission to scout the Territory. On the 18th they were near the Mescalero reservation, rather expecting the hostiles up that way. Their chief packer, named Burgess,[12] with a companion went over to a little Mexican town for fresh vegetables; and, coming back, ran right into **Nanay's** band, which had sneaked up from Mexico. Both packers were seriously wounded and their mules stolen. They reached Guilfoyle's camp on the 19th, and an hour later Bennett took up the

---

[7]It was not Victorio who struck at Arizona seeking his dependents, and who fought at Ash Creek, but his son, Washington, later killed in New Mexico. Victorio at the time was otherwise engaged.—Thrapp, *Victorio and the Mimbres Apaches.*

[8]Ineffectually, as it turned out. Buell and his massive expedition never caught sight of Victorio, nor came close to him, the Indian then being far to the east.

[9]Colonel (not General) Joaquin Terrazas (1829–1901) led a long and determined scout for the hostiles, commanding militia forces. Victorio, virtually out of ammunition, at length was surprised at Tres Castillos, Chihuahua, his band destroyed, and those not slain taken prisoner. Nana and a few others, separated from the main body, escaped to raid again.—Thrapp, *Victorio and the Mimbres Apaches.*

[10]Nana did not return to the United States until July of 1881. There was virtually no hostile action north of the Border during the first half of the year.

[11]Guilfoyle (1853–1921) was a West Pointer from Maryland who joined the 9th Cavalry in 1877 and won a brevet for gallantry in three actions against Nana this summer. He pursued the swift-moving Apaches relentlessly, sometimes caught up, but never could conclusively best them. Guilfoyle retired as a colonel in 1917.—Heitman; Cullum; Thrapp, *Dictionary of Frontier Characters.*

[12]This might have been John DeWitt Burgess, active in Apachería from at least 1867 until after the Apache wars had ceased, and from West Texas to Arizona. There is a photograph of him on horseback facing p. 168 of Vol. I of James H. McClintock's *Arizona.*

hostile trail with his scouts. Next day he caught up with Nanay at the Whitewater, where the hostiles had just killed, with horrible indignities, two men and a 16-year-old girl, all Mexicans. A running fight followed for sixteen miles, until Bennett's stock was worn out, and he had to wait for the pack-trains to come up. Three days later he again overhauled Nanay in the San Andre[s] mountains, and captured all the hostile camp outfit, killing two men, one of whom was shot by our old friend Chihuahua, then Bennett's sergeant of scouts. Thence the hostiles went up the Rio Grande into the San Mateo mountains, carrying three Mexican boys captive, and pursued by twenty-six Mexicans and ten Americans. While Bennett was crossing the river, these citizens camped in Red Rock cañon. They were just settling down when Nanay's forces opened fire from the surrounding rocks, in which they had all the time been hidden. They killed four and wounded three of the pursuers, the rest fleeing, as indeed they had to. Bennett cut across the north end of the San Mateo mountains, and at 4 p.m. the next day struck the hostiles at Monoco [Monica] springs. They were traveling then 100 miles a day, and never stopping to build a fire. Here again he captured their effects and killed one of their men. Between the river and the mountains the hostiles had killed five people and taken three prisoners. From Monoco they had a running fight with Bennett for 16 miles, and then escaped through the Dothill [Datil] mountains, where they killed six Mexican men and one woman, and in a little town beyond they killed a Mexican and captured 117 head of horses. Thence they whipped to the east, killing two men and a woman at Servilleta,[13] September 2d. Here Bennett struck the first troops he had seen since he first started in pursuit, 41 days before—Capt. [Charles] Parker,[14] of the Ninth cavalry. Bennett had held his scouts ten days beyond their time; and in the forty-one days' pursuit had traveled 1247 miles. He now turned over the trail to Capt. Parker, and went to Fort Wingate [N.M.], arriving there with but

[13]There was a place named Sevilleta about 20 miles north of Socorro, New Mexico, and this may be the place Bennett meant; it does not show on modern maps but is mentioned in T.M. Pearce, *New Mexico Place Names* (Albuquerque, University of New Mexico Press, 1965).

[14]Parker, who served as a captain with Illinois forces during the Civil War, joined the 9th Cavalry afterward. He saw some service against Victorio and is sometimes confused with Henry K. Parker, a notable chief of scouts during that campaign. Charles Parker died in 1890.—Heitman.

[15]This summary of Bennett's supposed role in chasing Nana is not confirmed by Harold Miller, who has done the best published summary of the great Apache foray ("Nana's Raid of 1881," *Password, op. cit.*). In a later communication he said he had no recollection of Bennett from his research, but added that this was not conclusive. Bennett's summary, on the other hand, gives obvious evidence of his complete familiarity with the raid. Some of his details are found nowhere else; thus there is no reason to doubt that he had an important role in the pursuit, and his summary of activities may well be entirely correct.

nineteen of the fifty-two pack and saddle animals with which he had started on the pursuit. All the rest had worn out and had to be shot.[15]

This is but a sample bit of the many hard lines of service Mr. Bennett has seen in his fifteen years of scouting. He has served under [August V.] Kautz [commander of the Department of Arizona], Hatch, [brevet Major General David S.] Stanley, [probably Colonel Luther P.] Bradley, Crook, Mackenzie and others, and has good words for all of them.

Mr. Bennett will probably go to San Diego to-morrow, to visit his father, who is now 84 years old. After his short vacation he will return to headquarters in Arizona, and may undoubtedly be heard from there again.

> [*The above article was not signed by Lummis,*
> *but in all probability*
> *was written by him.*]

Los Angeles Times, May 9, 1886:

THE APACHES.

Interesting and Startling News from the Frontier.

O.D. Crocker, President of the Copper King Mining Company, located in the Cananea mountains, Sonora, arrived yesterday from the mines via Nogales, and proceeded to the Sierra Madre Villa [near Pasadena, California]. He made the trip to Nogales in the night to dodge the Apaches. On the way he received a note by courier from Lieut. [*sic*] Lebo, stating that the Indians are making for the Cananea mountains. Capt. [Charles A.P.] Hatfield[16] has a command of forty-two men in a position to ambush the Indians on the south side of the Cananeas, in which direction Lieut. Lebo is driving them. The White Horse Cavalry, sixty-two men, have a position ten miles north, guarding the only other pass to the Sierra Madre. The Indians are supposed not to be aware of any troops in the Cananea mountains, and Capt. Lebo, in pursuit, heard with surprise from Mr. Crocker that he was driving the Indians into two companies of United States troops. The distance from the scene of Lieut. Lebo's fight with the Apaches to the Cananeas, where Capt. Hatfield awaits them is about fifty-six miles. Mr. Crocker further states that Lieut. Lebo reported that three

---

[16]An Alabama-born West Pointer, Hatfield (1850–1931) came from the West Virginia Hatfield clan of the McCoy feud fame and "must have inherited at least one of the habits of his ancestors for on one occasion, when he had trouble with another officer, he invited him to 'shoot it out' before breakfast." The duel was averted. Despite such impetuosity, Hatfield was a good officer. He had some artistic talent and was a particular friend of John Bigelow, author of "On the Bloody Trail of Geronimo."—Patch.

of the Apaches fell, and he believes three were killed outright instead of two, and several were wounded. The fight was so spirited that Lieut. Lebo had no opportunity to carry off the dead, but from the blood left on the field of battle he was assured that several Apaches were wounded, certainly five or six Indians.[17]

It will be remembered that when Chihuahua was brought to Fort Bowie he told Gen. Crook that the famous Mangus was farming down in Mexico, and had not been with Geronimo for many months. In fact, it was impossible to learn just where Mangus was. Mr. Crocker declares that Mangus was scalped and cremated seven weeks ago at Hostead's ranch in the Guagulpe [Guasachuca] Mountains, sixty-four miles southeast of Cananeas, and sixteen miles from the Sonora river.[18] In February the United States troops lost the hostiles' trail in the Guagulpe Mountains, through a light fall of snow. The following month the same hostiles raided several ranches on the Sonora river, taking all the supplies they could find; thence they went to Hostead's lower ranch and raided that, driving the inmates to the mountains. A few days later, at Hostead's upper ranch, at 5 a.m., the Indians were seen coming up the cañon by a vaquero, who alarmed the household. Hostead and four companions, singly, repaired to the corral, away from the house. The Indians shortly after made a break from cover towards the corral, and arriving at some trees, near the corral, Mangus raised a pair of pearl opera glasses to his eyes, and while in the act of scrutinizing his surroundings was shot dead by Hostead. This shot was followed by a volley from Hostead's companions. The Indians endeavored to retake the body, but being unsuccessful decamped. Chief Mangus was 6 feet 1 inch in height, well advanced in years, wore spectacles and was well clad. An accurate description was taken of the body, and his apparel, scalp and ear preserved. The body, however, was burned. The trophies have since been identified. Thirty-eight dollars and fifty cents in American money was found wound in his hair in a knot on the top of his head. Mangus's wife is one of the captives sent to Florida, and she did not know of her husband's death. Mr. Crocker saw all of the clothing, which was ornamented with a flattened silver ornament on the right breast, the same as is worn by Geronimo.

IN HOT PURSUIT.

General Miles has got his plans in shape, and evidently intends to push the campaign into the enemy's country. Captain [Henry W.] Lawton,[19] the herculean officer who was Mackenzie's favorite, crossed the line yesterday at Nogales with 80 picked infantry and 20 Apache scouts. He is well provided with all the necessaries, and has also ample means to purchase whatever he may need in Mexico. He is given *carte blanche* as to expenditures. His orders are to follow Geronimo's trail even though it takes him to the City of Mexico. He may be gone six months or more, no limit of time having been made. Captain Lawton also takes with him some members of the Signal Service Corps. The heliograph signal service is found to work very well, although it is not yet brought to perfection. A message was sent in this manner, a few days ago, a distance of 150 miles in four hours.

> [*The above story is unsigned, but it appears*
> *to have obvious Lummis touches,*
> *although it is dubious if he wrote it all*].

[17]Thomas Coverly Lebo (1842–1910) at this time was a captain, not a lieutenant, and already had compiled an outstanding record as an Indian fighter and officer of "good judgment, energy and conspicuous gallantry." He had his fight May 3 in the "Penito Mountains," of Sonora. These may have been in the vicinity of Las Pentas, just south of the border, 30 miles east of Nogales and 24 miles north of Cananea, Sonora. Hatfield would have a fight May 15 in the "Santa Cruz Mountains" of Sonora, possib;y east of Santa Cruz, which lies 21 miles southeast of Nogales.—Thrapp, *Dictionary of Frontier Characters.*

[18]This rumor was unfounded. See Note 20, Chapter 2.

[19]Lawton was one of the more famous soldiers to emerge from the Geronimo War and went on to a distinguished military career, cut short when he was killed in the battle of San Mateo, in the Philippines, December 19, 1899.—Lane, *Chasing Geronimo.*

CHAPTER XIV

Los Angeles Times, May 9, 1886:

CRAWFORD.

The Authentic Story of His Murder.

NOW NO LONGER TO BE SUPPRESSED.

*An Affair Which Merits a National Demand
for Satisfaction—Lieut. Maus's Statement.*

IF ANY CAP-SHEAF had been needed to the monumental fact that this United States is the only civilized nation on the face of the globe which does not protect its own citizens when away from home, the Crawford murder fills the bill. Here was not only an American citizen, but an American officer, wearing the uniform of his country, vested with its authority and sent in its name across the Mexican line, in accordance with treaty stipulations. His command of United States troops was attacked by a Mexican force for the apparent purpose of robbery; and as gallant a captain as the army ever had was shot down in absolute wantonness by those Sonorian banditti. So long as there was a chance of capturing and looting our camp, the Mexicans kept up the fight. When they were whipped and driven back, they suddenly discovered that it was a "mistake." The barefaced apology was readily been caught up by the American press, and it is astonishing to find that almost the only mention of the matter in the papers of the United States has been in the line of salving-over the murder. Had poor Crawford been a Briton, a German, or anything else but an American, his death would have been atoned ere this by blood or indemnity; but as it is, his countrymen content themsevles with apologies for his murderers. The damning facts which prove the plea of error a threadbare lie have been emasculated and suppressed. But murder will out. I will lay before you the facts; and you may choose between the word of those Mexican semi-military banditti—of the most ignorant, mendacious and brutal class among quasi-civilized nations—and that of intelligent and honorable officers of the Army of the United States.

Capt. Emmet Crawford and his four companies of Apache scouts—White Mountains and Chiricahuas—organized November 9, 1885; left Fort Bowie, November 29; and on the 11th of December crossed the Mexican line in pursuit of Geronimo and his band. The preliminary operations need not be detailed; but after a long and incredibly laborious pursuit the scouts located the hostile camp, February [January] 9, 1886, among the savage recesses of the Espinosa del Diablo, the raggedest mountains in all mountainous Mexico.[1]

Leaving the trail ten miles from where the camp was supposed to be, the command made a cautious night march; and just before dawn, on the 10th, was within a few hundred yards of the hostiles. The surprise would have been a success, probably, but some of the enemy's burros began braying, and the hostiles took to the rocks in the darkness. The story will now be carried on in the statements of those concerned. The appended map, though not on scale, will give an idea of the situation and the operations. I have shown only the hills directly concerned in the two engagements, but it should be understood that the whole country there is frightfully mountainous.

LIEUT. MAUS'S STATEMENT.

First Lieut. Marion P. Maus, First Infantry, upon whom the command devolved at Crawford's death, is a faithful, intelligent and honorable officer. He says:

"Just before daylight on the 10th we surprised the hostile camp [See point A on the accompanying map.] We were trying to surround them, but it is a tamalpais[2] country, and a few rolling stones started their burros to braying, and the hostiles, taking the alarm, hid in the rocks and opened fire on us. Our scouts responded. It was still perfectly dark, and we couldn't see their camp, but were guided only by the flash of a rifle here and there. No one was hurt on our side, but we found several pools of blood in the hostile camp. They escaped into the mountains before day light. We got their whole camp outfit— blankets, dried meat, roasted mescal. etc., to the amount of several wagon loads—with thirty horses and mules. We had been marching steadily since 11 a.m. the day before, and were worn out. Capt. Crawford sent back for his pack-grain of eleven mules, and we camped just beyond the bronco [hostile] camp [at B] to rest. Next morning, the 11th, we were awakened about daybreak by our Indian scouts, who called out that Mexican soldiers were coming. As nearly as I can fix the time, it was about daybreak. I could recognize a person probably 100 yards away. Immediately a severe fire was opened by the Mexicans. Lieut. Shipp and I, with Mr. [Tom] Horn,[3] chief of

[1]The Espinoza del Diablo, it will be recalled, is part of the main Sierra Madre, lying just north of the Aros (Haros) River and south of the Rio de Satachic, on the border of Sonora and Chihuahua.

[2]Probably means *malpais*: bad country.

[3]Tom Horn (1860–1903) is one of the best-known and, paradoxically, one of the most enigmatic figures of the early West. Born in Missouri, he reached Arizona at some undetermined time (his "autobiography" is not reliable on that and other matters) and probably by 1882 had become a packer with the army. He continued in that capacity until on this expedition he had his first experience as chief of scouts. He performed bravely and well during the rest of the Geronimo War. Horn became a top steer roper, often winning match contests. During the Spanish-American War he was a packer in Cuba; his career afterward is highly controversial. He was hanged at Cheyenne, Wyoming, for murder.

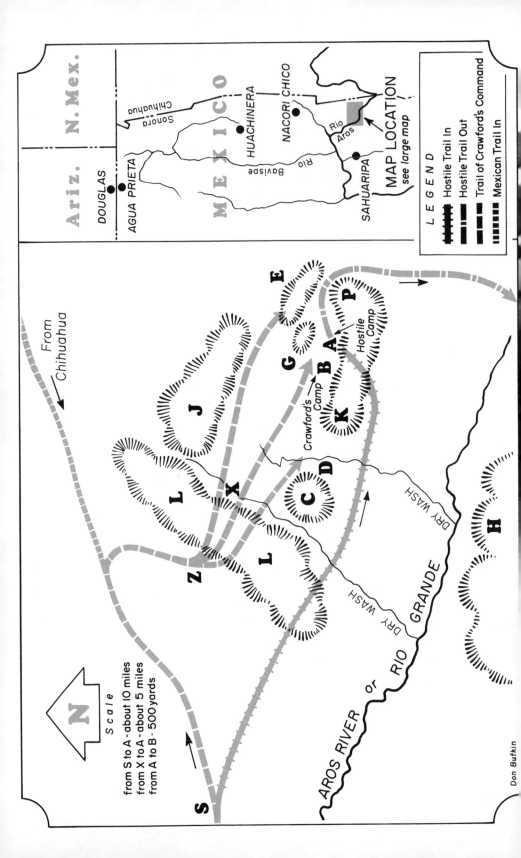

scouts, ran out to stop any attack. The firing was so heavy as to stop us before we got to our scouts, some sixty yards away, and we sought shelter. The Mexicans had taken a high position about 200 yards north [at G], and were firing down into us. Our scouts called out in broken English who we were; and I in Spanish kept calling, 'We are American soldiers. Don't shoot. We are friends.' All of us kept shouting, thinking the Mexicans had attacked us by mistake. We didn't want to kill them, supposing they were friends and hadn't recognized us. At our command, our scouts did not fire at all until the Mexicans had been firing about ten minutes; and even then there were but a few shots, and these from scouts so far away [toward P] that we couldn't stop them. Our command being concealed, in Indian fashion, the Mexicans had no idea how many we were, and by their initial actions showed they thought we were very weak numerically. Please remember that important fact. They were so near, and we were calling so loudly and constantly that they stopped after at least fifteen or twenty minutes' firing. When the firing ceased a party of nine of them came out directly to where we were; and fourteen or fifteen more circled around further to our front and right, while a third party, farther off, moved toward our right flank, to get possession of a high point [P] which commanded our position (They had not reached that point before firing recommenced). The nine first mentioned came up to within thirty yards of where our scouts were in the rocks. Crawford and I, from different parts of the rocks, walked out to talk to them. I was within six feet of the Mexican captain, and said to him in Spanish: 'Don't you see we are American soldiers? Look at my uniform and the captain's.' He looked very nervous and frightened, and with another man kept saying: 'Si! si! si! [yes, yes] *No tieres! No tieres!* [Don't shoot! Don't shoot!]. It was plain enough why he was nervous. He was now so close to our camp that he could see among the rocks the heads and rifles of 50 or 60 scouts. He moved off, continually saying *'No tieres!'* and seemed to be trying to get behind a little hill. At this time, Capt. Crawford, seeing the necessity of having no shot fired on either side, said to me, 'Maus, for God's sake go back there to the scouts and see that not a shot is fired!' I turned and passed in front of him, and had gone about two-thirds of the 30 yards between us and the rocks, when I heard a single shot behind me, instantly followed by a heavy volley. That first shot was distinct to me above all else. It seemed like a death knell. It flashed across me, 'Great God! These men are killing each other for nothing!' and on the heels of the thought came the volley. All of us sought cover. Crawford had come back from the open ground, got upon a rock and turned back his face to the Mexicans, when he was shot through the forehead. The scouts say that the first shot was the one that hit him. It was fired by one of the party of nine Mexicans with whom we had been talking. As soon as Crawford fell, our scouts opened fire on those nine Mexicans, killing four instantly and wounding four more. All the other Mexicans in sight then sheltered themselves, and some kept so close that they did not even fire again. Our firing then was very heavy, and they *had* to 'get.' Within a minute and half after I turned and saw Crawford lying on the ground, a scout called to me 'Look there! Look there!' I turned and saw the party of Mexicans, already mentioned, trying to get to the high point [P] on our right. I told some of the scouts to go quick and keep them out. Four or five scouts were already there; and with those whom I sent over, they drove that party of Mexicans a mile away. They had been trying to sneak to that commanding position during the cessation of firing, but hadn't time. The cessation had been about ten minutes. Within half or three-quarters of an hour after the second firing began, the Mexicans had been driven out of range, and their shots became very straggling. They had, so they said, 154 men. We had 79 Apache scouts and six white men present. Several of the Mexicans were still within easy hearing, and Mr. Horn had been yelling to them all the time in Spanish; and *after the firing had ceased and their attack had failed,* I heard one of them calling, 'Come over and talk! Come over!' Mr. Horn started out toward them. I was going to call him back, but he is a courageous fellow, and went on about 250 yards, to where some of them were lying, and commenced to talk to them. I followed him. They said it was 'all a mistake,' and they 'were very sorry,' 'what a great pity' [*que lástima*], and so on. I was glad to have the murderous thing stop, and, without reflection, in the hurry and excitement, accepted their explanation. They said we had killed their Captain, Corador [Mauricio Corredor], and also the next in command. The third in command, who talked to me, was a sergeant named Santana Perez [Santa Ana Perez]. There was another sergeant named Rancon. These two did the talking. They told me they were after the hostiles. I told them explicitly what we had done—how we had captured the camp and animals of the broncos—and they understood it perfectly. They told me they had been following the trail of these Indians all the way from Chihuahua, and had traveled nineteen days. This is not true. We had followed the hostile trail for eight days from west to east, while the Mexicans came from the northeast and never got nearer than a mile from the trail. They did strike our trail four or five miles [at X] from where they attacked us and followed it. They told me they were in dreadful condition, worn out and without food. They asked me to give them something to eat, appealing piteously. Wherever you meet these Mexican volunteers, they always are begging for something, generally food. They asked me to let our surgeon [Thomas B. Davis] come over and keep their wounded from bleeding to death. If I would send him over, and let them have half a dozen of our captured horses to carry their wounded they would leave, as they wanted to go right home. I had many reasons for being anxious to get rid of them. Overtures of surrender had already been made the day before by Natchez, who sent in two women, one after the other, to treat with us. I believed that as we had got hold of their all except their persons and their arms; there was a good chance that the hostiles would have surrendered unconditionally; but they would never do it so long as the Mexicans were around, for they always suspect Mexican treachery. They have had abundant cause to in the past. Furthermore our rations were nearly out, our one belt apiece of ammunition nearly exhausted. It was raining, and the country—almost impassable for pack-trains, anyhow—was hourly getting worse and worse. When our train would get in from Nacori was a matter of the greatest uncertainty. If the river got too high, it couldn't reach us at all. Two of our wounded (Captain Crawford and a scout) had to be carried by hand, so we would have to move very slowly. We were over 200 miles' march from the line. I promised to do what I could for the Mexicans. The doctor was not with us then, but he got in that day; and, as soon as he had attended to our wounded, he went over and cared for theirs. That day, at their request, and to get rid of them, we exchanged papers. The one Sergt. Perez gave me was to this effect, that they 'had met us unknowingly, and attacked us, on the Haros river, January 11th, and fired upon us, not turning back.' That it was dark, so that they couldn't see at the time, and that they 'fought us by mistake.' My paper was to the effect that we had met them unknowingly on the Haros river, being in camp at the time, and couldn't see perfectly when the first attack was made. (It *was* a little foggy when they first fired upon us; though, as I have said, I could have recognized a person at 100 yards; but when they killed Crawford it was perfectly light.)

The doctor told me it could not injure our wounded to move them, and I commenced to make litters. Next morning, the 12th, I sent over six of the captured horses to within 100 yards of where the Mexicans were [D], and went on with my preparations for leaving. The Mexicans refused to come down that 100 yards and get the horses, but demanded that we bring them clear up. Mr. Horn and the interpreter, who had taken the horses over, refused to go any nearer, and told the Mexicans to talk with me. We were within half an hour of starting (about noon), when some scouts told me that several of the captured horses had wandered off to the hill [LL] beyond the Mexicans. I sent Concepción, a Mexican interpreter over after them. When he got over to where the Mexicans were, they stopped him and asked why he hadn't brought the horses to them. He answered: 'Why didn't you come and take them when we brought them? I'm going after my horses—you go and talk to Lieut. Maus.' He was then going on, when the Mexican sergeant turned to his men and said: 'Keep this interpreter, and don't let him escape. He shall die here with us.' Thereupon Concepción sat down. They told him to call me over, and he did. He was exactly where some of the Mexicans were when they first opened fire on us, about 200 yards from me, and I could hear him distinctly calling that they wouldn't let him go, that there was some trouble, and he wanted me to come over—that it would be all right for me to come. I saw that he was in trouble, and I couldn't leave him there in any fix. Horn was wounded [slightly, in the arm], and I was the only one left who could converse in Spanish. I went half-way across the ravine [between K and D] and said to them: 'Come down here if you want to talk to me. What have you got to say?' They answered: 'Come up here; come up. We want to talk with you about those horses, and see if we can't fix it up. Don't fear anything from us—we are friends.' They kept on protesting friendship, and I went up on the hill, where about a dozen of them were standing. It was raining, and they proposed that we should go under a projecting rock, a few yards off. I went with them. As soon as we got around a little corner, I saw the position. Full fifty Mexicans were lying up against the cliff, with their guns. They said: 'Sit down. We want to fix up this business. We want those horses.' I said: 'I sent over the horses—why didn't you receive them?' They answered: 'You didn't bring them up to us.' Their manner was such as to make me think my position a ticklish one. They were a tough-looking set, and acted very menacingly. I thought it best to take no notice of this, and said, in an off-hand way, that I would go to my camp and send over the horses. They said: 'No, send this man [Concepción] after them.' Then I saw the whole game, and perceived that they had trapped me. So I gave a note to Concepción for Lieut. Shipp, asking that he send over six of the captured horses. Concepción then spoke up and said to me: 'Captain, there are many of our horses over here. Let this people take them.' Then I said to the Mexicans: 'Take six of our horses that are over here.' They declared there were none. Concepción said, rather angrily, that there *were*. I told the Mexicans to send out and see if he wasn't right. Some of them went out and brought in a roan mare, saying there were no others. Then Concepción went over to our camp, and drove five of the captured horses over to me. The stock was pretty rough, but there were six (with the mare), and that was all I had promised. The Mexicans, however, refused to accept them, saying: 'We won't have these— they're no use.' I said: 'Well, I'll go over to camp, and send you six serviceable horses that you *will* take.' I got up and started to go, but they stopped me. How? They jumped in front of me and all around me, a large number of them, saying, 'No! no!' in a menacing manner. I said: 'Do you mean to tell me that I can't leave here and go to my own camp?' They told me very plainly that that was just what they meant. Then they commenced to ask me what right I had in Mexico, what papers I had, etc. I had no papers. Capt. Crawford had had the papers, but they were with the pack-train at Nacori. I told them they should know that I had a right there; that, under the existing treaty, troops of either nation could cross the line in pursuit of the hostiles—and that, as Mexican soldiers, they must know this fact. They told me they knew no such thing. Their manner grew more and more unfriendly, and their talk among themselves was threatening. Concepción said those around him demanded that all the Americans be brought over to them; and said that they wanted our mules, and that our doctor must go with them to care for their wounded. I finally told the Mexicans I had no papers, but that Capt. Crawford had a letter from the Presidente of Sahuaripa. I sent Concepción over to get this letter from Capt. Crawford's coat. He went, and told our folks that I was a prisoner, and that the Mexicans proposed to go with us wherever we went. Our scouts got mad, and said they would take to the mountains before they would go with the Mexicans, whom they could not trust. They jumped up and began to strip for a fight, shaking their guns and calling the Mexicans names. The Mexicans said: 'Look at that. They are enemies of ours.' I answered: 'If you detain me here, I can't control my men. You'd better let me go to them.' The Mexicans understood perfectly that I was an American officer. I told them so, time after time, and bade them beware how they treated me. I had on my uniform and shoulder straps, to which I called their attention. But they refused to let me go. Meantime Concepción had got back with the letter, which they read. It was a very friendly missive from the Presidente of Sahuaripa, recognizing Captain Crawford as an American officer, in command of United States troops; telling him where he would probably find the trail of the hostiles, and what depredations they had committed. My captors read this letter over, but seemed to think it amounted to nothing. They demanded that I should give them six of our [government] mules for their wounded; I demurred, but they insisted; and at last, seeing it the only chance to escape a disastrous fight, I promised to give them six government mules. Even then they would not let me go until our scouts made that hostile demonstration. Standing up on the rocks so, the strength of our command was then first appreciated by the Mexicans, who were evidently impressed. At last they let me go, having extorted my solemn promise to give them the mules, and sending three or four of their number with me to get the mules. They held Concepción a hostage until I sent the mules over and demanded his release. My scouts said that they had not fired on the Mexicans, knowing that if they did, Concepción and I would be instantly killed. These Indians, taken from a barbarous race and given a trust, kept it faithfully. Fighting for a government whose flag was no protection to them in that alleged friendly country, they showed more honor and forbearance than the Mexicans who are called civilized. They had endured hunger, cold, and fatigue; had taken us through places where no American troops ever went before. They had been marching over those terrific mountains day and night, in rain and cold, without fires, and, at the time we attacked the bronco camp, 48 hours without food. No one could have been more faithful.

"When I got back to our camp, at last, and had quieted the scouts, I sent the six mules over to the Mexicans, with a note saying this was all I could do for them; that we wouldn't camp with them as they demanded, and that if they annoyed us further I should call on the Mexican authorities. They sent no receipt for the mules till I sent a note calling for one; and further informing them that we were going to leave; that they must not follow us, and that if they did they would cause trouble. I moved out my command under cover, leaving a company of scouts to hold the ground until we reached a commanding position, where they rejoined us. When we had gone about three miles we saw the Mexicans starting off in the opposite direction, and we saw them no more. The hostiles had gathered at H., and watched our fight with the Mexicans, thereafter departing.

"We had no more trouble with Mexicans till we reached Baserac, about twelve days later, on our march home. Before getting to the town we learned that the Presidente had organized a force of fifty men there, and ninety at Bavispe, to march up to the line with us and see that we went out of the country. I sent word to him that if he sent any troops for such a purpose I should consider them as enemies. They should neither camp nor march with us; and he should see the wisdom of keeping them at home, after the grave complications which had arisen. At Baserac we met Major [Emilio] Kosterlitzky,[4] a regular Mexican officer, of the Eleventh cavalry. He understood the situation, and protested, with me, against the sending of Mexican forces with us; but it was with the greatest difficulty that these forces were prevented, both there and at Bavispe, from going with us. Notwithstanding the fact that it is a breach of the Mexican laws to sell liquor to the Indians, mescal was brought through the bushes to our scouts, both at Baserac and Bavispe; and despite our most strenuous efforts, over half our scouts got intoxicated. It took the utmost exertion to ward off trouble between them and the Mexicans, but we succeeded. After leaving Bavispe we had no more trouble.

"As to the killing of Capt. Crawford, it could not possibly have been a mistake. Their first firing *could* have been an error, but not what followed. Underestimating our strength, they attacked us to rob. The killing of Crawford was deliberate assassination."

The statements of Lieut. Shipp and the rest will appear in another letter.

CHAS. F. LUMMIS

[4]Kosterlitzky (1853–1928) was one of the most colorful Borderland figures. Born in Moscow, he entered the Russian Navy, deserted ship in Venezuela, and made his way to Mexico, where as a *rurales* officer he came to widespread note. A Diaz man, he fled across the border in the course of the Mexican Revolution and settled at Los Angeles, where he performed valuable services for the Federal Bureau of Investigation and where he died. The best treatment of his life is Cornelius C. Smith Jr., *Emilio Kosterlitzky: Eagle of Sonora and the Southwest Border* (Glendale, California, Arthur H. Clark Co., 1970).

Nana, renegade Apache chief. Courtesy the National Archives.

The death of Captain Emmet Crawford, who was mortally wounded by Mexican irregulars in January, 1886, is here portrayed in an artist's conception for Miles' book, *Personal Recollections*. From *Personal Recollections*, by Nelson A. Miles (Chicago, The Werner Company, 1897).

Captain (then First Lieutenant) Marion P. Maus, who extracted the Crawford command from Mexico after Crawford, its commander, was wounded mortally by Mexican irregulars. Maus won a Medal of Honor for this work. This illustration appears in *Personal Recollections,* by Nelson A. Miles (Chicago, The Werner Company, 1897).

Crawford's Apache scouts, with whom he penetrated into the Sierra Madre, only to lose his life to Mexican irregulars. Photograph courtesy the Arizona Historical Society.

(*Facing page*) Captain Emmet Crawford. U.S. Signal Corps photograph, courtesy the National Archives.

Tom Horn, chief of scouts in the Sierra Madre during the Crawford campaign.

Nacori Chico, Sonora, as it looked at the turn of the century, probably not much different from its appearance in 1886. Crawford was buried here temporarily after he had been shot down by Mexican irregulars in the Sierra Madre. It always was important in Mexican-Apache affairs. Photograph courtesy Arizona Historical Foundation.

Brigadier General George Crook, in command of the Department of Arizona. This photograph, taken in 1885, shows him mounted on his mule Apache in an arroyo near Fort Bowie, Arizona. Courtesy Arizona Historical Society.

to them: 'Don't shoot! we are American soldiers!' They kept answering: 'Si! Si!' {yes, yes}, but were all the time edging around until they got behind a tree. Then they opened fire without warning. Mr. Horn jumped down off his rock, and grabbed his left arm. I asked him if he was hurt. He said yes. [The ball passed through the fleshy part of his arm]. The Mexicans who shot at him were about twenty-five yards away. We were both in full view. I had on a brown hunting-coat, with the blue army trousers; Mr. Horn a civilian's suit. As soon as we had sheltered ourselves in the rocks, after Horn was shot, the scouts called out that Capt. Crawford was killed. I went over to where he lay, about twenty yards away. He was on his back behind the rock upon which he had been standing. A silk handkerchief lay on the wound, which was an inch above his left eyebrow. A piece of his brain as big as my two fingers lay on the rock, and his hat was on the ground. He had on a full uniform, and a brown government field-hat when he stood there. It was impossbile to mistake him for an Indian. The Indians never have any hair on their faces; but Capt. Crawford had a large moustache and chin beard. I stepped off the distance, afterward, from where he lay to the spot where the men stood who shot him. Their position was definitely marked by a little tree. *The distance was twenty-seven paces.* My paces average about 30 inches. As soon as Crawford fell, our scouts opened fire, and so did the Mexicans on the surrounding hills. We had heard there were regular Mexican troops coming from Sahuaripa, and had supposed these were they. The Mexicans kept up this second firing about an hour, though we were calling to them all the time. The air was perfectly still; and many of them were within 300 yards of us all the time. Some of them were trying to get down a gully and up to a high point [P] to surround us; but a company of Chiricahua scouts held the point, and did most of the firing from our side. We told them not to kill any more of the Mexicans than they could help, but to keep them out of the rocks. I got a gun and kept it by me, but didn't shoot at all, nor did any of the white people in our command. *After the Mexicans had been driven completely back, several times, they began to answer Mr. Horn's constant calling.* He got up on a rock and talked to them. They asked him to come out. He went out to them without any weapons, shortly followed by Lieut. Maus, also unarmed. The scouts watched like hawks, but I kept them quiet. While Maus and Horn were over there, Dr. Davis and the pack-train came in. The Doctor went at once to Capt. Crawford, and pronounced the wound mortal. Lieut. Maus was in the Mexican camp three-quarters of an hour or an hour. Next day he went over again on some busines about the horses, and was gone about the same length of time. He sent Concepción over to me with a note for a letter Capt. Crawford had had from the Presidente of Sahuaripa, and telling me the Mexicans wouldn't let him go. I sent over the desired paper as soon as I could find it. Presently Lieut. Maus returned, with two Mexicans, and said he had had to agree to give them six government mules. We sent them over. We had already, that morning, sent over to them five captured ponies, which they would not receive. Next day we moved out without annoyance, and never saw those Mexicans again. I have thought the matter over a great deal, and carefully gone over the whole ground where they were and where we were. It is absolutely impossible that they could have been mistaken. All six of us white men were in full view of them when they shot Crawford. He had his full uniform on; Lieut. Maus a blue army overcoat and blue trousers; I my partial uniform; the hospital steward a full soldier's uniform. Mr. Horn had plain clothing, but he was standing on a rock in full view, with his hat in his hand. His hair is very light and short and his complexion very light. Many of the scouts had on their army blouses and chevrons. When Capt. Crawford was shot I could see perfectly the faces of those who shot him. It was then broad daylight. The man who shot Mr. Horn was Miguel Carador [Mauricio Corredor], the Mexican commander. He was immediately killed. The scout, 'Dutchy' killed the man who shot Capt. Crawford. The Mexicans were dressed in the ordinary raggedness of their class. I heard no orders given among them. They claim in their papers that they had been following the hostile trail seventeen days. I know that is false for we went back over the trail, and found that they had come in from the northeast, while the trail came from the west. They never got on it at all.

"Capt. Crawford lived seven days and four hours after he was shot. We carried him on a litter for two days, eight scouts bearing the litter at a time, all over terribly rough country. It was raining all the time, but we kept him dry. After we met [Henry] Daly's pack-train from Nacori with more rations, we fixed one end of the litter to a mule, six scouts carrying the other end. The country was horrible. There was no level ground there, and the utmost we could make was six or seven miles a day. He died on the 18th, on the road about 30 miles from Nacori. On the 17th he seemed to be somewhat conscious, and nodded his head feebly when we came up to him. But he never spoke a word after he was shot. We carried him to Nacori, and buried him there with great difficulty. There were only four boards to be found in town. I have known Capt. Crawford for a long time, and never saw a man who knew so much about the Indians. He was personally acquainted with all the hostiles and all the scouts, and knew all their traits. He was an ideal Indian-fighter, and an ideal man." [This is concurred in by all who knew poor Crawford.]

MR. HORN'S STATEMENT.

Thomas Horn, [Spanish-English] interpreter and Chief of Scouts, is a manly and reliable Missourian. He says:

"On the morning of January 11th we were camped in the rocks close to the camp whence we had jumped the broncos on the preceding morning. About daybreak we saw the Mexicans coming, about 500 yards off, along a ridge near this hill [J]. They were straggling along in no sort of order, save that about twenty were ahead. The rest stopped on that hill [J] 400 to 700 yards from our camp. They opened fire when they got within 400 to 600 yards. When they began firing, our scouts called out that we were American troops, and yelled to Capt. Crawford and me that the assailants were Mexicans. Capt. Crawford called to me to go out and speak to the Mexicans. I went out and got upon a rock, probably thirty-five yards from him. Most of the Mexicans stopped on the ridge [J] but several came up close to where we were. Three of them passed me within thirty yards, going toward Capt. Crawford. I called out to them in Spanish: 'We are American soldiers, and the Indians here are soldiers, too.' The first three passed me without stopping; but the fourth stopped when I said the same to him. When he stopped, he drew a bead on me at once. I was unarmed. I called to him in Spanish: 'Lower your rifle! We are American soldiers!' He lowered his rifle about a second, and smiled. Then he took a quick aim at me and fired, hitting me in the left arm. Then firing on both sides began, and he was killed at once. I had on a cloth civilian's suit. The firing then lasted nearly an hour. I kept yelling to them in Spanish: 'We are American soldiers. Why don't you answer?' They made no reply, *till we had driven them off*. Then they answered, 'What do you want?' I told them to stop firing a minute, and I'd come over and tell them what we wanted. They said: 'All right. Put down your arms and come over.' I took off my cartridge belt, laid down the gun I had picked up, and went over. They said: 'What have you been doing here?' I answered: 'I came over to tell you that we are American troops.' They asked where was our captain; and I answered that he was badly wounded—probably killed. Then one of them, an old man, said: 'Our captain is killed, too. Here is his gun (shaking a Winchester in his hand) and *I'd like to kill a Gringo* [white man] *with the same gun.*" I remember this expression positively. One of the others told the old man to be quiet. I called Lieut. Maus over, and he talked with the commanding officer of the Mexicans. The officer said it was all a mistake—they didn't know we were Americans. He and Lieut. Maus agreed to exchange papers to the effect that it was a mistake. *Then the Mexicans demanded all the horses and burros we had captured from the hostiles.* They claimed that they were citizens of towns below, and that the Indians had stolen the stock from them. Lieut. Maus told them they couldn't have the animals, which belonged to the scouts, who had captured them from the broncos in the fight of the day before. Then the Mexicans said *part* of the stock was theirs; and wanted all rounded up, so that they could pick out theirs. They said they needed five or six to take their wounded home on. Lieut. Maus told them he'd think the matter over. Then he and I went back to camp where our pack-mules were just coming in. We had had nothing to eat for forty-eight hours, after traveling day and night, with two fights thrown in. In the evening we went over to the Mexican camp again, and exchanged the papers as agreed. They asked: 'What's the matter with your scouts, that they don't come around to see us?' Lieut. Maus said he didn't want the scouts in their camp, because the scouts didn't like the Mexicans. They said: *'If your Indians want to fight let them come out in this little flat, and we'll give them some more of it.'* Lieut. Maus told them we didn't come down there to fight Mexicans, and wouldn't unless we had to. They said: 'All right.' Then we returned to our camp, the Mexican commander saying that he wanted to be friends with the Americans, and that we could come and go and visit his camp as much as we liked. Next day, Lieut. Maus went over there at their request, and soon sent back a note for some horses, which they refused to take when sent. Later, he sent a note for six mules. Concepción, who fetched the note, told me Lieut. Maus was a prisoner. He also told the Indians of the matter. They said, 'We'd better fight it out right here, then; for if we give them six mules, they'll send back for all the rest.' They began to fix themselves in the rocks and to strip off all their clothes, as they always do for a fight. They were calling to one another, and dancing around where the Mexicans could see and hear them. In a few minutes Lieut. Maus came over with two Mexicans, by whom he sent back six mules. Next day about noon we left, the Mexicans being still in camp.

"They might have made a mistake in the first few volleys, which were fired from a distance; but the rest was no mistake. Those who came close up to us, within twenty-five yards or less—I saw Lieut. Shipp pace off the distance— didn't fire at the Indians at all. Capt. Crawford and I were the only ones hit by that fire. If they didn't see the Indians, why should they fire at us Americans? And if they did see the Indians, why should they fire on the Americans first? They evidently thought that if they could kill the Americans they could run off the scouts, and get the plunder."

DR. DAVIS'S STATEMENT.

Dr. T.B. Davis, surgeon of the battalion and brother of the late [brevet Major] Gen. Jeff C. Davis, says:

"During the fight I was up on a hill across a ravine; and got to camp about two hours after Crawford was shot. I found him lying in the mud behind the rock upon which he must have been in full view. He had been waving a handkerchief which he borrowed from hospital steward Nemick. He had on his uniform blouse and pants. I took the blouse off him myself. Next day the Mexicans got hold of Maus and held him for ransom. That makes me confident that they attacked our camp to rob it. thinking our force was small. When I went over to care for their wounded, after the fight, they showed me four dead and four wounded. The Mexicans had no sign of uniform. Some had blankets, and some only cotton clothes. Some were barefoot, some had moccasins, some sandals and some shoes."

Hospital-Steward [Frank J.] Nemick[1] has been ordered to his regiment elsewhere. His statement, however, fully coincides with those above, and was included in Lieut. Maus's official report

GEN. CROOK'S OPINION.

Brigadier-General Geo. Crook read the dispatch I showed him, stating that President Diaz (in his message to the Mexican Congress) holds that the murder of Crawford was a mistake. Gen. Crook said:

"The Mexican authorities do all in their power to assist us in hunting down the renegade Apaches. Those Mexicans who attacked Crawford were not regular troops, but simply banditti, after scalp-money and plunder. It was no mistake. They knew who it was. No one ever heard of an Indian or a Mexican making such a mistake. American troops might do it readily—but Mexican or Indian, never! The only mistake they made was in our numbers. They didn't know how strong our force was, and thought they could 'get away with it.' If the Mexicans hadn't attacked Crawford just then, I think this whole Apache business would have been settled then and there. Crawford knew the hostiles, and they would have surrendered to him when they wouldn't to any of the others. But they wouldn't have anything to do with the Mexicans, whom they have good reasons for distrusting."[2]

There have been, beside the Crawford assassination, several other occasions on which our forces were seriously troubled by Mexicans—as when Lieut. Elliott, on his way with a pack-train for Capt. Crawford last summer, was jailed at El Via, though he showed his papers as an American officer; and when Lieut. McDonald and his eighteen scouts, in 1882, were kept four days in the Casas Grandes jail; after proving their identity. But these are trivial matters beside the wanton slaying of a brave American officer whose blood it is apparently desired to hide with apologetic sand.

CHAS. F. LUMMIS.

---

[1]Nemick was a private in Troop G of the 4th Cavalry and at this time was a hospital steward, second class.

[2]For the official summary of the Crawford affair, which includes the statements above and others, see "Papers Relating to Foreign Relations of the United States (Mexico)," *House Executive Document* 1, 49th Congress, 2nd Session, Serial 2460, 570–691.

AFTERWORD

The End of the "War"

FOR REASONS HE DOES NOT CLEARLY DEFINE, Lummis never accompanied a scout column into Mexico as he had hoped to do, and for that he might have been grateful. All that remained of the campaign was an infinity of aimless wandering through the next to impassable mountains of Sonora in pursuit of wraiths who could not be caught up with by white soldiers, much less brought to battle.

Miles's "determined pursuits by white commands" were unproductive exercises in human endurance; as is well known they did little to achieve a conclusion of the Geronimo campaign. Rather it was Gatewood and the Apache scouts, brought once more to fulfill their historic function, who made contact with the hostiles and persuaded them to surrender in late August of this year, 1886. Geronimo had moved north to the vicinity of Fronteras, where he had met Mexican officials, ostensibly to arrange for a surrender but more likely, perhaps, to assuage his thirst for alcohol and to obtain desired supplies. The Mexicans always believed he fully intended to give up. In fact, many persons today are convinced that Geronimo actually did surrender to Mexicans first, to the Anglo commands afterward.[1]

In any event, by early September the surrender had been brought about. Shortly afterward, the last considerable body of erstwhile hostiles was shipped to Florida, and the formal Apache wars of the Southwest were at long last finished.

[1]See Dan L. Thrapp, "Geronimo's Mysterious Surrender," Los Angeles Corral of the Westerners, *Brand Book 13* (1969), 16–35 for the Mexican viewpoint of the surrender of Geronimo and the Anglo role in that event.

Los Angeles Times, June 11, 1886:

CROOK'S REPORT

Of His Last Apache Indian Campaign

AN INTERESTING PIECE OF HISTORY

*Which Has Never Been Published—Causes of the Present
Outbreak and Beginning of the Campaign.*

I HAVE HAD THE GOOD FORTUNE to secure a copy of General Crook's final report from the Department of Arizona. It has come to me entirely without knowledge on the General's part. The document is full of interest for many reasons. It is the authoritative statement of the causes and the operations of the present outbreak; and though the TIMES has already foreshadowed some of its conclusions, the document now sees light for the first time. It is doubly interesting in the present juncture. It is now two months since General Crook was relieved of the command of the Department of Arizona. He had then sent four-fifths of the renegades to a Florida prison. There were but 34 hostiles out—of whom 14 were squaws—and these were worn out and sought nothing but refuge in some inaccessible place. There had been no depredation in the United States for many months, and comparatively few in Mexico. Crook was removed. The confidence which the Reservation Indians had felt, that they would not be at the mercy of the white ruffians of the Territory during his rule, was now replaced by doubt as to what the new man would do. The renegades learned of the change about as soon as anyone, you may be sure. Instead of the dreaded "Chief-with-the-brown-clothes," who knew all their haunts and all their tricks of warfare, there was now a new commander, of whose there-untried prowess they felt no fear. The Apache has a hideous humor in his composition, and it seems almost as if they renegades had been simply laying themselves out to show Miles's braggart friends—his friends, mind; *Miles* is no braggart—just how it was themselves. In this two months there has been more bloodshed than in the preceding two years under Crook; and the hostiles, flitting from rocky range to range, do a digito-nasal dance for the benefit of their futile pursuers. You keep hearing how this and that band of "Rangers" (legitimate successors to the Tombstone Toughs) has killed three or four renegades; but somehow they never bring in a head for the handsome reward. There chances to be a very efficient check on those prevaricators. Every Apache is tagged and numbered, and a full description of him is kept in a big book. This fact leads to the final discomfiture of the warriors who claim to have slain thus and so many hostiles. If we were to believe the dispatches, the whole reservation is on the verge of the war path, and the raiders have already been largely reinforced. This is probably further Territorial fiction; but, if true, it simply militates against the very point the venomous novelists are still trying to make—viz., further damnation of Crook's policy. There was no such dissatisfaction, no such rebellion, while Crook was there. The Territory is also illustrating itself by two propositions; one that the Apaches must all be removed from Arizona—in cool disregard of the fact that Arizona belongs to the Apaches, that they were forced to accept the small Reservation in lieu of the whole Territory, and that even the Reservation has been thievishly stolen from and cut down five times to fill the pockets of grasping settlers. The other proposition is the old assassin plan to march on the Reservation and butcher the women and children who cannot get away. There is precious little danger of that, however. Arizona would sooner storm that deadly fastness with its mouth than in fact.

There is an impression that unprecedented means are now being taken to catch the hostiles. Such is not the case. With the exception of a reward offered by the citizens, no more is being done than was done when Geronimo broke out a year ago, or when Ulzanna made his bloody raid last fall. The great difference is that under the new regime the value of advertising is understood. Miles is doing all he can, and is doing it well. I hope he will "get there"—and those remarks are not intended as any reflection upon him. But put on your thinking-cap and tell me what Arizona has gained in the last two months.

And now to Gen. Crook's report.

"HEADQUARTERS, DEP'T OF ARIZONA, IN THE FIELD, FORT BOWIE, A.T., April 10, 1886

Adjutant General, Division of the Pacific, Presidio of San Francisco, California,
SIR:—I have the honor to submit the following report of the operation of the troops under my command in the pursuit of hostile Chiricahuas.

"My first information of impending trouble was a telegram received on the afternoon of May 17, 1885; and before a reply could be sent, the wires between Fort Apache and San Carlos were cut. The next afternoon I was informed that Geronimo, Nanay, Mangus, Natchez and Chihuahua, with a considerable party, had left their camp just after dark the preceding evening. Within a few days the exact number of renegades was fixed at thirty-four men, eight well grown boys and ninety-two women and children. I learned that on May 15th Lieut. Britton Davis, 3d Cav., sent a telegraphic dispatch to men, which I did not see until months afterwards. Had this telegram reached me, I feel morally certain that the troubles would have been settled without an outbreak. Troubles of minor importance were constantly occurring on the Reservation, which were quieted down by the officers in charge reporting them to me and receiving my instructions. There probably would not have been as much danger or difficulty in managing the matter reported in Lieut. Davis's telegram as there was in quelling the disturbances started by Ka-e-te-na in March and June, 1884, in which Bonito was implicated, or those in July, 1885. I have not on hand papers relating to the management of those difficulties; but in each instance methods were employed suitable to the special emergency. This trouble arose from a tizwin drunk; and in order to shield the guilty parties, all the prominent chiefs drank of this liquor, and in a body went to Lieut. Davis and informed him of the fact, thinking that in this way all would escape punishment. Lieut. Davis told them that in a matter of such importance he could not take action himself, but would report the whole occurrence to me for my directions. Lieut. Davis wrote the telegram in the presence of the Indians, told them what he had written, and said that he should act in accordance with my personal instructions, and would notify them what my orders were when received. A messenger was immediately sent with this dispatch to Fort Apache, but no reply was received. The Indians waited until dark, and again assembled the next day, but receiving no reply became alarmed, and doubtless concluded that I was making preparations to seize the whole of them and punish them as I had Ka-e-te-na. This idea, I afterward learned, had been put in their heads by ill-disposed persons on the Reservation. A matter of this kind, while it must be managed with exceeding caution, must be settled promptly—without giving time for their suspicious imaginations to work upon their fears. Delay is a fatal error. For further details, attention is invited to the report of Lieut. Davis, which accompanies this paper. [I shall add this to the General's report, marked "Appendix A."—LUM]. [It was not so added, ED.].

"Within an hour after the renegades left their camp on Turkey Creek, two groups of the Fourth Cavalry, under command of Capt. Allen Smith,[1] and a party of White Mountain and Chiricahua scouts, under Lieuts. Gatewood and Davis, left Fort Apache in pursuit. But such was the rapidity of their flight that it was impossible to overtake them. It subsequently transpired that they traveled nearly 120 miles before stopping for rest or food. [Captain Smith's report of the immediate pursuit I have not at hand—L.]

"As soon as the departure of the Indians was known, troops were immediately put in motion to overtake or intercept them. Captain [Francis E.] Pierce,[2] with a

---

[1]Allen Smith (1849–1927) was born at Fort Marion, St. Augustine, Florida, to which the erstwhile hostiles would be shipped in 1886. He was the son of a Major General and a cadet at the U.S. Navel Academy, Annapolis, from 1863 to 1866, then became an infantry officer and later served in the cavalry. He became a Brigadier General and retired to Spokane, Washington, in 1905.—Heitman; *Who Was Who.*

[2]Francis E. Pierce (1833–1896) was born in New York City and amassed a fine Civil War record emerging as a brevet Brigadier General of Volunteers. He then served on the frontier as a junior officer; it was his mistake that led to the Geronimo outbreak in May of 1885, his action based upon faulty advice given him by Al Sieber. Thomas Cruse, prominent in Apacheria for serval years, called Pierce "one of the best 'Indian men' we ever had." He was named agent for the Sioux at Pine Ridge, South Dakota, shortly after the Wounded Knee action, and died from injuries suffered from a fall.—Thrapp, *Al Sieber, Chief of Scouts*; William H. Powell, *Powell's Records of Living Officers of the United States Army* (Philadelphia, L. R. Hamersly & Co., 1890).

party of scouts from San Carlos, moved toward Ash Peak. The commanding officer at Fort Thomas, without waiting for orders, sent two companies of cavalry toward Clifton. The commanding officer at Fort Grant was ordered to send all his available force—five companies of cavalry—toward the Gila, with orders to cut the hostile trail if possible, and to pursue vigorously, regardless of Departmental or National lines. The commanding officer at Fort Bowie was directed to ambush his cavalry at proper points in the Stein's Peak range, which had been a favorite trail in former years. The commanding officer at Fort Huachuca was instructed to send three troops to Guadalupe Cañon, and to scout the whole country in that vicinity. Information was sent to the commanding officer, District of New Mexico, of the departure of the Indians and the movements of my troops; and every effort was made to warn citizens, at all points within reach of their danger.

"On the 28th of May, my information indicating that the Indians had gone into the Black Range, New Mexico, I left my headquarters and proceeded to Fort Bayard, whence I could more intelligently direct the movement of troops from my department. The whole country north, east and west of Bayard was filled with troops. No less than 20 troops of cavalry and more than 100 Indian scouts were moved in every direction, to either intercept or follow the trails of the hostiles. But with the exception of the capture of a few of their animals by the Indian scouts under Chatto, and a slight skirmish with their rear-guard by the troops from Apache under Capt. Smith on May 22d, in which three of his command were wounded— the hostiles were not even caught sight of by the troops, and finally crossed into Mexico about June 10th.

"In the twenty-three days from the outbreak until the Indians crossed into Mexico, every possible effort was made by the troops, which were pushed to the limit of endurance of men and animals, but without result other than to drive the Indians out of the Black Range and Mogollons, and also to save the lives, probably, of many ranchmen and prospectors. It must be remembered that the two years of peace had enormously increased the business interests of Arizona and the contiguous portions of New Mexico. Cattle and horse ranches had been established wherever a mountain stream afforded the necessary water. Thus the hostiles found food and transportation in every valley; whereas the pursuing troops were limited to the horses they were riding and to the rations carried on their pack trains. The vigor of the pursuit may be understood from the fact that more than 150 horses and mules were found on the different trails—worn out and killed, or abandoned by the Indians in their flight.

"As soon as it became evident that the hostiles were moving south, I proceeded to Deming, June 5, and began preparations to follow them into Mexico.

"Captain Emmet Crawford, Third cavalry, who had reported to me, was ordered with the battalion of scouts which had been operating in the country about the old Warm Spring Reservation, to go to Separ by rail, and thence to move with a troop of cavalry to the south end of the Animas Valley, with the hope that the hostiles might cross into Mexico through the Guadalupe Mountains. The result showed that the main body of the hostiles crossed the Line on the west of the Mule Mountains, though a small party surprised a camp of the 4th Cav. in Guadalupe Cañon, guarded by seven enlisted men, of whom they killed four. Another small party crossed the Line near Lake Palomas. Lieut. Davis—who, with sixty White Mountain and Chiricahua scouts, had been following the trail of the hostiles as rapidly as possible, was ordered to report to Capt. Crawford. On the 11th of June the combined force—consisting of ninety-two scouts and Troop A, 6th Cav., followed the hostiles into the Sierra Madre. On the 9th of June a telegram [attached to the original report, but not in my hands—L.] was received from Lieutenant-General Sheridan, informing me that I was authorized to enlist 200 additional scouts; and that the Cheyenne pack-train had been ordered to report to me. [This pack-train was brought down and fitted out by Chief of Transportation Thomas Moore, of whom I have already told you—L.] I was also directed to establish my headquarters at a point either on or near the S.P.R.R. I accordingly proceeded to Fort Bowie. Directions were immediately issued for the enlistment of the new scouts—100 at San Carlos, and the remainder at Fort Apache. As soon as the 100 at Apache were enlisted, Lieut. Gatewood was ordered in command, to thoroughly scout the Mogollons and the Black Range, in order to determine definitely whether any of the hostiles were remaining in that region, as was persistently reported. This movement delayed his arrival at Bowie about twenty days; and though it was found, as I had expected, that there were no hostiles in New Mexico, I did not deem it advisable to organize a second expedition for service in Mexico until this fact was definitely established.

"As soon as necessary preparations could be completed, Capt. Wirt Davis, Fourth Cavalry, with a command consisting of his own cavalry troop and 100 Indian scouts, with pack-trains carrying sixty days' rations, was ordered into Mexico. [I have already given you a detailed report of his expedition.—L.]. My plan of operations was as follows: that the commands of Capts. Davis and Crawford should thoroughly scout the Sierra Madre and the adjoining mountain ranges, and endeavor to surprise the hostile camps in Mexico; at the same time I would so station troops along the border as to prevent, if possible, the return of the renegades to the United States when they should be driven out of Mexico by the commands operating in the mountains. With this in view, I placed a troop of cavalry at every water hole along the border from the Patagonia mountains to the Río Grande. With each troop I stationed a detachment of Indian scouts, with sufficient pack-mules to carry at least ten days' rations for the command. Orders were given to conceal the troops, and to keep the country between the different camps constantly patrolled. I also established a second line nearer the railroad, as reserves to the first line. In order to insure prompt supplies to the commands operating in Mexico, I established a depot at Lang's ranch, at the south end of the Animas valley, near the boundary. In New Mexico reserve camps were also established at such points as seemed to offer the best facilities for rapid movement and successful pursuit in the event that the renegades should return to the United States. Four troops of the Eighth Cavalry, from Texas, having reported to Gen. [Luther P.] Bradley,[3] were stationed north of

[3]Luther P. Bradley (1822–1910) was born at New Haven, Connecticut, and served through the Civil War, emerging a Brigadier General of Volunteers. He was a Colonel of the 13th Infantry at the time Crook mentions him and would retire by December, 1886, living afterward at Tacoma, Washington.—Ezra J. Warner, *Generals in Blue: Lives of the Union Commanders* (Baton Rouge, Louisiana State University Press, 1964).

the railroad, at points which were deemed most available for pursuit of the Indians, and most likely to afford protection to life and property. [That this was a much more thorough disposition of troops than the present, no military man will deny—L.]

"On the 23d of June Capt. Crawford's scouts under Chatto struck Chihuahua's band in the Bavispe mountains, northeast of Oputo; but owing to the position occupied by the hostiles, their camp could not be surrounded, and in the fight which ensued, they escaped. Fifteen women and children, a number of horses and a considerable amount of plunder were captured."

I will give the rest of this interesting report in a later installment.

LUM.

Los Angeles Times, June 27, 1886:

CROOK'S REPORT

Of His Latest Anti-Apache Campaign.

AN INTERESTING PIECE OF HISTORY

Never Before Published—Operations in Arizona, New Mexico and Old Mexico—Assassination of Crawford.

[SECOND INSTALLMENT]

Gen. Crook's report goes on as follows:

"Capt. Davis (Fourth Cavalry) crossed into Mexico, July 13, and after much severe fatigue succeeded in locating the camp of a band of hostiles under Geronimo in the Sierra Madre, a little northeast of Nacori. The camp was attacked by a picked detail of seventy-three scouts, under Lieut. Day, Ninth Cavalry; and, though the surprise was complete, it is now believed that the only hostiles killed were one squaw and two boys. Everything in the camp was captured, with fifteen women and children. [Nut for the long-range mouth warriors of Arizona, and their sanctum-campaigner abettors elsewhere, who gloat over the silly allegation that the scouts never were known to fight hostiles.—L.] A copy of Capt. Davis's report of this affair was forwarded at the time. Captains Davis and Crawford continued scouting in the mountains, and their commands endured uncomplainingly almost incredible hardships and fatigues. But the hostiles were so continually on the alert that no other engagement was had until, on September 22d, Capt. Davis again struck a band of about twenty hostiles in the Terez mountains. In the fight that ensued, one of the scouts was killed. [Nut No. 2 for the above mouth-warriors—L.] The hostiles having been driven out of Mexico by the scouts, crossed into the United States through Guadalupe cañon, within a few miles of a camp of two troops of cavalry, about daylight on the morning of September 28th. They were closely followed by both Captains Davis and Crawford. It being evident that the hostiles intended to raid the White Mountain Reservation, or go into the Mogollons or the Black Range, New Mexico, dispositions were made to prevent this. Cavalry was directed from different points by converging routes toward the Gila. Troops were established in positions to prevent the hostiles crossing the San Simon valley into the Stein's Peak range; and others were placed along the railroad where they would be available for instant transportation by rail to threatened points. The scouts followed the hostiles, and several troops of cavalry were moved to points where it was thought possible to ambush them. The renegades took the roughest possible trails over the Chiricahua mountains, and twice endeavored to cross the San Simon valley, but each time were frightened back into the Chiricahuas by seeing the dust of moving columns, or discovering their trails across the valley. They then crossed the Sulphur Springs valley by night into the Dragoons, whither they were followed by Crawford's scouts. They fled through this range, back into the valley, and south toward the Mule mountains. Here their trail suddenly turned sharp to the east and went back into the Chiricahuas, Crawford and his scouts following persistently. The stock of the hostiles was by this time worn out, and though they had gathered all that was possible along the way they were finally absolutely [unhorsed] and the troops were in such position that it seemed probable that the entire band would be captured or killed. But just at this juncture they succeeded in remounting themselves with the best stock in the country; and finding that it would be impossible to get north of the railroad, they returned to Mexico. Capt. [Charles D.] Viele,[4] Tenth Cavalry, followed them, with two troops, as far as Ascencion, Chihuahua; from which point, further pursuit being useless, he returned with his jaded command to his camp in Cave Cañon.

"The remounting of the hostiles was in this instance particularly exasperating. The cattlemen of the San Simon had gathered in White Tail Cañon, on the east side of the Chiricahuas, for the beginning of their fall round-up. In spite of warnings which they received the evening before, that hostiles on foot had been seen in the vicinity, they lariated their cow ponies, the best stock in the country, around a ranch in which they all slept. In the morning all their stock except two or three were gone, and the hostiles had secured about thirty of the finest horses in Arizona. This is not an isolated instance. Several times before and since parties of hostiles have been dismounted by persistent pursuit, and escaped in the same way by securing remounts—and this, too, in spite of constant warning and importunities to ranchmen to secure their stock. The Indians acted as if they could take stock with perfect impunity. At one time they took a quantity of horses from a corral belonging to the Sulphur Springs Cattle Company under circumstances that make it evident that several men who were in the ranch-house knew what was going on. Although there were only three Indians in the raiding party, no attempt was made

---

[4]Charles D. Viele (1841–1916) was born at Albany, New York and served through the Civil War and on frontier duty afterward, participating in numerous Indian campaigns and scouting operations. He became a Brigadier General of Volunteers in 1898 and served in Cuba during the Spanish American War—*Who Was Who*.

to prevent the stock from being taken. At another time, early in June a party of Indians, numbering perhaps a dozen men and 40 or 50 squaws and children, drove up and shot several beeves within a mile of the largest ranch in Arizona, in broad daylight. There were 20 cowboys on the ranch at the time, all fully armed, and yet the Indians went into camp, cooked the meat, and finally left, sometime during the night. During all this time, not the slightest attempt was made to interfere with them, or even to give information to the troops.

"The hostiles having returned to Mexico the troops were sent back to their field stations. The scouts having been constantly on the march since the beginning of operations, and the terms of service of many of them having expired, it was thought best to discharge them and enlist others; and, while the new commands were being organized, to refit and reorganize, as thoroughly as possible, the pack-trains, which were by this time almost worn out. The new commands were fitted out as soon as possible; and, on November 27th, Captain Davis again started into Mexico. Captain Crawford was detained by a fruitless pursuit of a raiding party under Ulzanna, and did not cross the line until almost two weeks later.

"The raid of the party of 11 hostiles who succeeded in eluding the troops on the line and went up into New Mexico by the Lake Palomas trail early in November, will not be reported at length, as the raid was mentioned in a special report, dated January 11, 1886 (copy attached, marked E.) and the operations incident thereto treated at length. [I have not secured this special report.—L.] It is mentioned as showing the dangers and difficulties to be contended against from small parties. During the period of about four weeks this band traveled probably not less than 1,200 miles, killed 38 people, captured and wore out 250 head of horses; and, though twice dismounted, succeeded in crossing back into Mexico with the loss of but one man, who was killed by loyal Indians, whose camp they attacked near Fort Apache. At one time it seemed probable that the band would be captured, but the refusal of a party of forty Navajo scouts, under Lieut. [Albert B] Scott,[5] Thirteenth Infantry, to follow the trail, although supported by a troop of cavalry, caused the pursuite to be abandoned, owing to the inability of the troops to follow the trail of the hostiles in the mountains, and a severe storm of snow and rain coming on, which lasted three days and obliterated all trails as fast as made, so that the raiders were enabled to choose their own route into Mexico.

"For details connected with the movements of Capt. Davis's command in Mexico, I respectively refer to the attached reports of this officer, marked F and G, respectively. [I have not this document, either—L.] The first expedition of Capt. Crawford is well covered by the report of Lieut. Britton Davis, attached, marked H. [This I am also unable to procure at present.—L.] It is to be regretted that the death of Capt. Crawford—at a time when there is reason to believe that, had he lived, he would have received the unconditional surrender of Geronimo, Natchez and their

---

[5] Albert B. Scott (c. 1858–1906), a West Pointer from Texas was an infantry officer who capped his frontier service with gallantry in Cuba during the Spanish American War. He was wounded in the charge up San Juan Hill and died of the effects of his injuries less than a decade later.—Cullum.

bands—has prevented a detailed report of all the operations pertaining to his first expedition. The report of Lieut. Maus of the second expedition is attached, marked I. [This report I hope to have shortly. You have had the most important part—that detailing the assassination of Capt. Crawford by Mexicans.—L.] The detailed report of the engagement with the hostiles by Capt. Crawford's command, and the subsequent death of this gallant officer in an attack upon his command by a Mexican force, is fully traversed by attached reports of Lieut. Maus (marked K and L), who assumed command upon Crawford's death.

"From these reports it appears that on the 10th of January, Capt. Crawford—after an exceedingly difficult night march—attacked the hostile camp, near the Haros river, about sixty miles below Nacori. Though the attack did not result in the destruction of the hostiles, for the reasons shown, yet the scouts captured all the hostiles' stock and supplies of every description, and convinced them that they could never find a secure resting-place. The hostiles accordingly asked for a conference with Capt. Crawford, to take place the following morning. What the results of that conference would have been cannot, of course, be positively stated; but he was thoroughly known to all the Indians, and had their confidence; and it is believed that he was the only white man besides myself who could have induced the hostiles to surrender. Unfortunately the scouts—worn out by three days' incessant marching, for forty-eight hours without food; and, under the circumstances, fearing no attack by the hostiles, did not keep watch that night with their usual vigilance. The morning of the 11th, before full daylight, they were attacked in their sleep by 154 Mexicans. By the first volley three of the scouts were wounded before they could get the shelter of the rocks. By the exertions of their officers the firing was stopped, and every effort was made to explain to the Mexicans that they were attacking a friendly force. The Mexicans were told in Spanish that the Indians were American scouts, and that the officers in command were American officers. During this interval the Mexicans approached so near that their words could be distinguished. Capt. Crawford took his position on a rock, without arms, within easy speaking distance of them. He pointed out that he was in uniform. At the same time Mr. Horn, the interpreter, was explaining why they were. Suddenly, without warning, a Mexican, within about twenty or thirty yards of Capt. Crawford, raised his piece and fired. Crawford fell, shot through the brain. This shot seemed to be the agreed-upon signal, as at once the firing became general. Mr. Horn was shot through the arm; but though the firing lasted several minutes [over half an hour—L.], and was only stopped after the Mexicans had lost their commanding officer and the second in command, with at least two others killed and several wounded, not one of the scouts was touched. Had it not been for the exertions of Lieuts. Maus and Shipp and the two Chiefs-of-Scouts, who finally succeeded in stopping the firing of the scouts, many more of the attacking party would have been killed.

"In the light of the events of the day following—when Lieut. Maus was detained by force in the Mexican camp, and was only allowed to return [to his own camp] when his scouts began to strip for action, and upon his word of honor that he would send them [the Mexicans] six of his eleven pack mules—the conclusion reached by Lieut. Maus, that the Mexicans knew whom they were attacking, is almost controvertible. *The death of Captain Crawford was, in any event, an assassination.*"

This is plain, up-and-up talk from a man whose absolute truthfulness was never questioned, even by his most hydrophobic enemies. The conclusion of this interesting report, giving a graphic and full statement as to the surrender of Geronimo and the close of Crook's administration, must be reserved for another installment.

LUM.

Los Angeles Times, July 2, 1886:

CROOK'S REPORT

Of His Last Anti-Apache Campaign.

CONCLUSION OF HIS OPERATIONS¡

*Another Chapter of Unpublished History—From the
Conference in Sonora Up to the Miles Regime.*

The conclusion of Gen. Crook's report follows, and brings the Apache war down to the time when Gen. Miles took the reins:

"The day following the Mexican attack, the hostiles again asked to talk; but Crawford was insensible, practically dead (though his death did not actually occur until the eighth day), and the interview of the hostiles with Lieut. Maus resulted only in their expressing a wish for a conference with myself at a point indicated—a point about 25 miles south of San Bernardino. This conference took place in the Cañon de los Embudos on the 25th of March. I found the hostiles encamped on a rocky hill, surrounded by ravines and cañons through which they could escape to the higher peaks behind, in the event of an attack. They were in superb physical condition, armed to the teeth, and supplied with all the ammunition they could carry. In manner they were suspicious, and at the same time confident and self reliant. Lieut. Maus, with his battalion of scouts, was camped on lower ground, separated by a deep, rugged cañon from their position, and distant 500 or 600 yards. The hostiles refused to allow any nearer approach.

"I was conscious that in agreeing to meet them I was placing myself in a position similar to that in which Gen. Canby lost his life, and that any incident which might—with or without cause—excite their suspicions, would result in my death and probably that of some of the officers with me. I therefore endeavored to induce the hostiles to meet me within the United States, urging that the presence of white soldiers would prevent any attempt of the Mexicans to attack them. But no argument would move them. Into their hands I must trust myself, or back they would go to their mountain fastness. The report of this conference has been made in full [I sent you all the important points and speeches from Fort Bowie—L.] After my first talk, it seemed as if they would accept no terms other than that they be allowed to return to the reservation upon the *status ante bellum.* This I positively declined, and the conference broke up. That night I got emissaries into their camp, but the hostiles were so excited that they would listen to nothing. The friendly Indians whom I employed as emissaries, told me that they dared not even talk to the hostiles of surrendering. Geronimo told his people to keep their guns in their hands, and to be ready to shoot at a moment's notice. The friendly Indians said that the slightest circumstance which might look suspicious would be a signal for firing to begin, that the hostiles would kill all they could, and scatter in the mountains. Even after they surrendered to me, they did not relax their vigilance. They kept mounted and constantly on the watch; there were never more than five to eight of their men in our camp at one time; and even after the march northward began, the hostiles did not keep together, but scattered over the country in parties of two and three. At night they camped in the same way; and, had I desired it, it would have been an absolute impossibility to seize more than half a dozen of them. The remainder would have escaped, and our breach of faith would have prevented forever any possibility of any settlement with them.

"The last conference with the hostiles took place on the afternoon of March 27th. That night whisky or mescal was smuggled into their camp, and many of them were drunk. The next morning Chihuahua reported this fact, but told me they would all begin to move toward the border. In order to be in telegraphic communication with the War Department, I deemed it essential to return to Fort Bowie immediately, and I accordingly left the camp, leaving there my interpreters and the trusted Indians whom I had employed in the negotiations, with Lieut. Maus—who with his battalion of scouts, was to conduct the hostiles to Fort Bowie. The first day the command marched to the supply camp, about twelve miles south of San Bernardino; and the next day they camped at Smuggler's Springs, near the Border. Owing to the persistent sale of intoxicating liquor to the hostiles by a man named Tribolet [an unsavory character whom I have frequently referred to, and who is now visiting this city—L], I gave directions that his ranch should be so guarded that no liquor could be had by the Indians. It seems that from the time the scouts had been in camp, south of San Bernardino, this man had been selling to them—and boasted of the large amount of money he was making on the traffic. The exertions of Lieut. Maus and his officers did not prevent the hostiles from obtaining liquor from this same source on the night of the 29th, although the day before all the liquor which could be discovered by careful search (some fifteen gallons) has been destroyed.

"The night of the 29th the hostiles were apparently sober; and two dispatches from Lieut. Maus to me indicated that there would be no difficulty in continuing the march without trouble. But the Indians were in such a condition of mind that any remarks with reference to what would happen to them when they finally got into my power excited them. It is understood, and I believe, that such remarks were made by interested parties; and in consequence thereof, Geronimo and Natchez, with twenty bucks, stampeded during the night, taking with them two horses and one mule, fourteen women and two young boys. Several days afterward, two of the bucks returned. They said they were sleeping together, and heard their people leaving camp, whereupon—supposing something to be wrong—they left also. Next morning they concluded that there was no reason why they should leave, and they started back, voluntarily rejoining Lieutenant Maus about fifteen miles from this post.

"After the most careful inquiry, I am satisfied that no one in the camp, except those who left, knew anything about it until next morning; and it is probable that a number of those who left were frightened out at the last moment. Under the circumstances it would have been impossible to prevent their escaping. Lieut. Maus, with eighty scouts of his battalion, immediately started in pursuit. Capt. Dorst, Fourth Cavalry, with the First Battalion of Indian scouts, was also put upon the trail. [The result of these pursuits I have long since detailed—L.]

"The remainder of the prisoners arrived at Fort Bowie on April 2d, and on the 7th, in compliance with telegraphic instructions of the Secretary of War, left Bowie Station by train, under charge of First Lieut. J. R. Richards, Jr., Fourth Cavalry, under escort of a company of the Eighth Infantry, for Fort Marion, St. Augustine, Florida. They numbered 77—15 bucks, 33 women and 29 children.

"Among the warriors sent to Fort Marion are Chihuahua and his brother Ulzanna (who led the raid in November and December) and several others of almost equal prominence—the Indians say, the bravest and ablest of the hostiles. With the prisoners are two wives and three children of Geronimo; the family of Natchez, and families or relatives of all the hostiles who remain out. By this surrender the entering wedge has been well driven; and it is believed that there will be but little difficulty in obtaining the surrender of those who are still out. These are Geronimo and Natchez, with 18 men. Mangus and three men are also out; but where, is not known, he having separated from the other renegades in August last, since which time nothing has been heard of him. There is no evidence to show that his band has had any part in the outrages committed by the other renegades, and he will doubtless surrender when he can be communicated with.

"There were 34 men and 8 well-grown boys who left the Reservation in May last, and two small boys have since grown big enough to carry arms, making 44 in all. There are now 20 with Geronimo, and 4 with Mangus, total, 24.

"Before closing this report, I desire to express my appreciation of the conduct of the officers and men of my command during the dreary months they have been engaged in this discouraging and well-nigh hopeless task. Where all have done well, it seems invidious to mention individuals; but while my thanks are due to all, it seems proper to mention the names of Capt. Wirt Davis, Fourth Cavalry; the lamented Crawford, who sleeps in a soldier's grave; First Lieut. M. W. Day, Ninth Cavalry; First Lieut. Marion P. Maus, First Infantry; Lieut. Britton Davis, Third Cavalry; Lieuts. C. P. Elliott, R. D. Walsh and H. C. Benson,[6] Fourth Cavalry;

---

[6]Harry C. Benson (1857–1924) was born in Ohio and graduated from West Point, seeing some frontier service and reaching the rank of Colonel. He was something of a mathematician and won attention also as an ornithologist. After the turn of the century he performed duty at Sequoia, Yosemite and Yellowstone National Parks. He retired in 1915 but was recalled to active duty in 1917, serving until 1919. He died at San Francisco.—Cullum.

[7]Born in South Carolina, Finley became a second lieutenant in the 10th Cavalry in 1879. He rescued Colonel Benjamin H. Grierson in a hard fight against Victorio in 1880. After his Apache wars experience he became first lieutenant and died in 1894 of the effects of a Fort Custer, Montana, accident.

Leighton Finley[7] and F. E. Shipp, Tenth Cavalry; and S. L. Faison, First Infantry, who commanded expeditions or scout companies in Mexico, and bore uncomplainingly the almost incredible fatigues and privations, as well as the dangers, incident to those operations.

Very respectfully,
Your obedient servant,
GEORGE CROOK,
Brigadier General Commanding."

Thus ends this interesting and authentic chapter of the history of the Apache war—a chapter of which the TIMES is the only paper in the country to secure a copy.

LUM.

INDEX

INDEX